Botanical Research and Practices

Botanical Research and Practices

The Anti-Inflammatory Activities of Herbal Plants and Toxic Properties of Medicinal Plants
Wenli Sun, PhD, Mohamad H. Shahrajabian, PhD
and Mehdi Khoshkharam, PhD
2022 ISBN: 978-1-68507-909-3 (eBook)

Coriandrum sativum: **Origin, Uses and Nutrition**
Mamta Pujari, PhD, Bharat Kapoor, PhD,
Neeta Pande, PhD and Anju Joshi, PhD (Editors)
2022 ISBN: 978-1-68507-842-3 (Softcover)
2022 ISBN: 978-1-68507-955-0 (eBook)

Oregano – The genus Origanum (Lamiaceae): Taxonomy, Cultivation, Chemistry, and Uses
Tuncay Dirmenci (Editor)
2021 ISBN: 978-1-68507-315-2 (Hardcover)
2021 ISBN: 978-1-68507-387-9 (eBook)

Punica granatum: **Cultivation, Properties and Health Benefits**
Rupesh K. Gautam, PhD and Smriti Parashar (Editors)
2021 ISBN: 978-1-53619-767-9 (Hardcover)
2021 ISBN: 978-1-53619-821-8 (eBook)

Elementary Botany
George Francis Atkinson (Editor)
2021 ISBN: 978-1-53619-448-7 (Hardcover)
2021 ISBN: 978-1-53619-518-7 (eBook)

More information about this series can be found at
https://novapublishers.com/product-category/series/botanical-research-and-practices/.

Dr. Mohammad Saleem Wani
Dr. Younas Rasheed Tantray
and Dr. Afaq Majid Wani

Ethnobotany, Phytochemistry, and Pharmacology of Woody Angiosperms of the Northwest Himalayas

Copyright © 2023 by Nova Science Publishers, Inc.
DOI: https://doi.org/10.52305/SOVA1458.

All rights reserved. No part of this book may be reproduced, stored in a retrieval system or transmitted in any form or by any means: electronic, electrostatic, magnetic, tape, mechanical photocopying, recording or otherwise without the written permission of the Publisher.

We have partnered with Copyright Clearance Center to make it easy for you to obtain permissions to reuse content from this publication. Please visit copyright.com and search by Title, ISBN, or ISSN.

For further questions about using the service on copyright.com, please contact:

Copyright Clearance Center
Phone: +1-(978) 750-8400   Fax: +1-(978) 750-4470   E-mail: info@copyright.com

NOTICE TO THE READER

The Publisher has taken reasonable care in the preparation of this book but makes no expressed or implied warranty of any kind and assumes no responsibility for any errors or omissions. No liability is assumed for incidental or consequential damages in connection with or arising out of information contained in this book. The Publisher shall not be liable for any special, consequential, or exemplary damages resulting, in whole or in part, from the readers' use of, or reliance upon, this material. Any parts of this book based on government reports are so indicated and copyright is claimed for those parts to the extent applicable to compilations of such works.

Independent verification should be sought for any data, advice or recommendations contained in this book. In addition, no responsibility is assumed by the Publisher for any injury and/or damage to persons or property arising from any methods, products, instructions, ideas or otherwise contained in this publication.

This publication is designed to provide accurate and authoritative information with regards to the subject matter covered herein. It is sold with the clear understanding that the Publisher is not engaged in rendering legal or any other professional services. If legal or any other expert assistance is required, the services of a competent person should be sought. FROM A DECLARATION OF PARTICIPANTS JOINTLY ADOPTED BY A COMMITTEE OF THE AMERICAN BAR ASSOCIATION AND A COMMITTEE OF PUBLISHERS.

Library of Congress Cataloging-in-Publication Data

ISBN: 979-8-88697-627-4

Published by Nova Science Publishers, Inc. † New York

Contents

Chapter 1	*Betula utilis* D. Don: Phytochemistry, Pharmacology, and Ethnomedicinal Uses	1
Chapter 2	*Corylus avellana* L.: Traditional Uses, Phytochemistry, and Pharmacology	9
Chapter 3	*Ficus carica* L.: Traditional Uses, Phytochemistry, and Pharmacology	23
Chapter 4	*Juglans regia* L.: Ethnobotany, Pharmacology and Phytochemistry	43
Chapter 5	*Morus alba* L.: Traditional Uses, Phytochemistry, and Pharmacology	57
Chapter 6	*Rhododendron arboreum* Sm.: Ethnobotany, Phytochemistry, and Pharmacology	75
Chapter 7	*Shorea robusta* Gaertn. f.: Ethnobotany, Phytochemistry, and Pharmacology	91
References		101
Index		135
Authors' Contact Information		141

Chapter 1

Betula utilis D. Don: Phytochemistry, Pharmacology, and Ethnomedicinal Uses

Abstract

The current chapter deals with the description, active constituents, therapeutic uses, and pharmacological activities of *B. utilis*. Traditionally, the plant has been used for many ethnobotanical purposes, including religious and medicinal purposes. Being an essential source of medicine, different plant parts of *B. utilis* must be screened effectively for making standardized herbal drugs. At the same time, the collection of this tree species needs to be governed for its sustained availability and protection of the delicate Himalayan ecosystem. Subsequent studies and research can be done on the plant for its numerous pharmacological activities. The literature for the present review has been collected from various scientific databases viz. Science Direct, Web of Science, Scopus, PubMed, etc. and from various research articles and books.

1. Introduction

Betula utilis D. Don., frequently called Himalayan birch, is a member of the family Betulaceae. There are 136 taxonomically identified species of genus *Betula*. The species is a deciduous tree attaining a height of 15-20m and grows in the altitudinal range of 2800 - 4511m (Wani et al., 2018b). The bark of the tree is shining, with white horizontal smooth lenticels. The outer bark is present in the form of layers and shows extensive shedding. The leaves are ovate acuminate, elliptic, and occasionally serrate. Birches often bloom at the comparatively early age of 10 to 15 years.

The species is wind-pollinated and monoecious in nature (De Jong, 1993; Lawesson, 2001). Male catkins develop in summer and pollinate female flowers in the next spring. Their pollen grains are widely distributed and travel long distances by air (Johnsson, 1974, Hjelmroos, 1991). The male catkins are cylinder-like and tightly intertwined scales are produced terminally on small, spur-like sideways branches and emerge with the leaves. Once the female catkins mature in pre-fall or pre-winter, they turn amber and woody in appearance.

Seeds develop in consequence of cross-fertilization due to the biochemical self-incompatibility mechanism (Hagman, 1972). At maturity, the seeds transform from greenish to light amber. With leaf burst in the spring, the Himalayan birch flourish and seeds develop in October to November. The wind plays a crucial part in dispersing and scattering seeds on the snowy ground up to 80 meters from the tree. Although there are many seeds in forest soils, these seeds are short-lived. After the second or third year, most seeds are non-viable. In sandy, loamy soils, the plant develops, but may thrive in degraded or wet soils.

The existence of this birch species rely upon sufficient water and nutrients (Carlton and Bazzaz, 1998). Birch is a long-lived species, that has lived for over 400 years (Bhattacharyya et al., 2006). In ecological terms, *B. utilis* have an important contribution in subalpine and high Himalayan regions and is the sole broadleaved tree species present at these elevations (Zobel and Singh, 1997). It assists to keep up the delicate Himalayan biodiversity by inspecting soil run-off and producing bio-shield for the remaining woodlands and sub-alpine grasslands underneath the tree line region (Kala, 1998).

In addition to its multiple environmental benefits, *B. utilis* has been recognized for numerous ethnomedicinal importance by various non-ethnic and ethnic groups residing in the Himalaya and beyond. *B. utilis* has a wide sub-zero resistance (Sakai and Larcher, 2012), that allows this species to establish a Himalayan tree line. It is a common pioneer species, ready to set up on new, barren ground. *B. utilis* is regarded as an indicator species for climate-driven tree line dynamics (Bobrowski et al., 2017). The actual extent of *B. utilis* is little and scattered. Besides, habitat loss, grazing and fuelwood cutting by common residents are factors that have led to decreasing ranges of *B. utilis* by 98% falling under criteria A of critically endangered species (Haq, 2012; Samant et al., 2007; Zaki et al., 2011). Thus, the declining population from the Himalayan region needs protection.

1.1. Taxonomical Description

- **Kingdom:** Plantae

- **Division:** Magnoliophyta
- **Class:** Mangoliopsida
- **Order:** Fagales
- **Family:** Betulaceae
- **Genus:** *Betula*
- **Species:** *utilis*
- **Synonyms:** The tree species is known under various synonyms such as *Betula utilis* var. *glandulifera* Regel; *Betula utilis* var. *jacquemontii* (Spach) H.J.P. Winkl.; *Betula utilis* subsp. *jacquemontii* (Spach) Kitam.; *Betula utilis* var. *latifolia* Regel; *Betula utilis* subsp. *occidentalis* Kitam.; *Betula utilis* var. *sinensis* (Franch.) H.J.P. Winkl.; *Betula utilis* var. *utilis*.

1.2. Vernacular Names

- **English:** Himalayan silver birch
- **Hindi**: Bhoj patra
- **Tamil**: Bhurjjamaram, Purchcham
- **Malayalam**: Bhujapatram, Bhurjjamaram
- **Telugu**: Bhujapatri
- **Kannada**: Bhuyapathra
- **Sanskrit**: Bahulavalkalah, Bahupata, Bhurjapatraka
- **Nepali**: Bhoj Patra

2. Distribution

The genus is restricted to the Northern Hemisphere, particularly to the cool and temperate zones. In its geographic range, it grows from 2,500 m and 4,300 m. The species *B. utilis* is distributed to various parts of the world viz., Afghanistan, Bhutan, China, India, Kazakhstan, Kyrgyzstan, Nepal, Pakistan, Tajikistan, Uzbekistan (Wani et al., 2020).

It forms the topmost area of the natural tree line zone in the Indian Himalaya, typically ranging from 3300±200 m in the northwest to 3800±200 m in the northeastern region. At low altitudes (3400 m), it is occasionally found, however at upper altitudes, it has thick and clubby stands in either pure mixes or usually with *Rhododendron campanulatum* undergrowth or with *Abies pindrow*, *A. spectabilis*, and sometime with *Junipers* (Zobel and Singh, 1997).

3. Traditional Use of *Betula utilis*

The epithet, *utilis*, means the number of benefits of this imperative tree species. *B. utilis* is regarded holy by Hindus and is employed in different spiritual ceremonies. The external bark of this species has been in use since ancient times for publishing literature, particularly in Sanskrit. The external bark of this tree is a very useful part, which is still obsolete for writing mantras. To protect someone, particularly children, from all misfortunes and to receive blessings an amulet containing the bark of *B. utilis*, written with mantras, is worn on them. This talisman is put around the throat or attached around the arm. It is likewise used to make various Vedic charms, named as yantras and various yantras are drawn over its paper bark.

The collection of paper bark to make such yantras can be achieved at certain specific days and times by considering a specific star (a planetary house in Hindu horoscopy). Such yantra is thought to give a good living and even wealth (Kala, 2018).

Pilgrimage is a religious activity in the Chamoli district of Uttarakhand carried out after every 12 years and is known as the Nanda Devi Raj Jat Yatra. During this function, a unique umbrella is made for the idol of the local goddess Nanda Devi. The cover of this umbrella is composed of papery bark of *B. utilis*. This umbrella is locally known as Nanda's chhatoli. The wood from *B. utilis* is utilized in havan. The bark is also commonly employed for packaging material and roofing.

In the medication of bladder infections and kidney and bladder stones, the leaves of the plant are effective. The wood for building and the foliage for fodder are used. The tree is most commonly used as a fuelwood which has caused the elimination or reduction of large areas of habitat. In local folk medicine, parts of the plant, together with fungal growth (bhurja-granthi), have also been used for a long time.

4. Chemical Compounds of *Betula utilis*

The bark of the *B. utilis* contains many chemical compounds like betulin, betulinic acid, methylbetulonate, methyl betulate, lupeol, lupenone, oleanolic acid, acetyloleanolic acid, sitosterol and leucocyanidin (Nadakarni, 1976). A new triterpene karachic acid extracted from the ethereal extracts of bark, was crystallized and it appears as a colourless crystalline solid (Khan, 1975). HPTLC (High-Performance Thin-Layer Chromatography) certified five marker compounds, namely betulin, oleanolicacid, β-sitosterol, lupeol and 3-acetyloleanolic acid. The marker compounds have been resolved by utilizing n-hexane: ethyl acetate (8: 2 v/v) as the solvent system on silica-gel plates. Studies have shown that among the different marker compounds, betulin is an important constituent with its natural abundance as 2.88 percent (Khan et al., 2012).

The main class of compounds reported from *B. utilis* bark are triterpenes. As per the review of literature, triterpenoids have been demonstrated as the source of drugs for various chronic and malignant diseases and there are ample evidence to suggest that either natural or synthetic triterpenoids have anticancer activity (Patlolla and Rao, 2012). Cancer is the second leading cause of death after heart disease. Betulin is the main compound in the bark of the *B. utilis* which has anticancer actions because it suppresses the growth of malignant melanoma along with liver and lung cancer (Kikuzaki and Nakatani, 1993).

Betulinic acid extracted from *B. utilis* bark has specific cytotoxicity against HIV-1 activity, anti-inflammatory, *in vitro* antimalarial effects and specific cytotoxicity against a wide range of cancer cell lines. Derivatives of betulinic acids have shown significantly greater efficacy and better therapeutic effects than current anti-HIV drugs in the clinical stage (Cichewicz and Kouzi, 2004).

Betulinic acid has undergone extensive studies due to its anti-HIV action and selective antitumor action against human melanoma cells. Some of the betulinic acid derivatives also exhibit elevated anti-HIV, anti-inflammatory, antiviral and hepato-protective activities (Dzubak et al., 2006; Urban et al., 2004). The Compounds separated from the bark, leaves, and roots of the plant are given in Table 1.

Table 1. Classes and names of compounds isolated
from the various parts of *B. utilis*

Sr. no.	Compound	Part of plant	References
1.	Betulin	Bark	(Sharma et al., 2010; Khan et al., 2012; Joshi et al., 2013; Mishra et al., 2016; Pandey et al., 2016; Šiman et al., 2016; Bhatia et al., 2017; Wani et al., 2018b)
2.	Betulinic acid		(Sharma et al., 2010; Mishra et al., 2016; Bhatia et al., 2017; Wani et al., 2018b)
3.	Lupeol		(Khan et al., 2012; Mishra et al., 2016; Pandey et al., 2016; Wani et al., 2018b)
4.	Oleanolic acid		(Khan et al., 2012; Mishra et al., 2016; Pandey et al., 2016; Wani et al., 2018b)
5.	3-acetyloleanolic acid		(Khan et al., 2012)
6.	β-sitosterol		(Khan et al., 2012; Pandey et al., 2016)
7.	Ursolic acid		(Mishra et al., 2016)
8.	β-amyrin		
9.	Betulin	Roots/Leaves	(Wani et al., 2018b)
10.	Betulinic acid	Roots/Leaves	
11.	Lupeol	Roots/Leaves	
12.	Oleanolic acid	Roots/Leaves	
13.	Karachic acid	Bark	(Khan, 1975)

5. Pharmacological Properties of *B. utilis*

Many biochemical compounds found in different portions of the *B. utilis* are used for various therapeutic purposes. The main biochemical compounds are pentacyclic triterpenes which are recognized as betulin, lupeol, betulinic acid, lupenone, β-sitosterol, methyl betulonate, and methyl betulate, while the minor compounds reported are oleanolic acid, ursolic acid and, betulinic aldehyde (Kumaraswamy et al., 2008).

B. utilis is also a source of essential oil, which includes geranic acid, seleneol, linenool, sesqui phellendrene, champacol and 1, 8-cineol, with strong antimicrobial activity against the fungi, *Candida albicans* and Gram (+) and Gram (-) human pathogenic bacteria namely *Staphylococcus aureus*, *Escherichia coli*, *Pseudomonas aeruginosa* (Pal et al., 2015). Several other pharmacological reports have been described from the different extracts of *B. utilis*. An overview of the reported biological activities is given in Table 2.

Table 2. An overview of the modern pharmacological assessments

S. no	Activity	Parts used	Extract	Organism used/Cell lines	Reference
1.	Cytotoxic	Bark	Methanol hexane, ethyl acetate, chloroform, water, and butanol	Lung (A549), Colon (DLD-1,), Breast (MCF-7, MDA-MB-468), Head and Neck (FaDu), Cervical (HeLa), Prostate (DU145), Ovary (SK-OV-3), Brain (A-172), Liver (PLC/PRF/5), Pancreas (MIAPaCa-2), Renal (786–0, Caki-1) Breast (MDA-MB-231) and Colon (SW 620)	(Mishra et al., 2016)
2.	Antibacterial	Bark	Petroleum ether, chloroform, methanol, ethanol, and water	*Citrobacter* sp. *Escherichia coli* MTCC 443 *Klebsiella pneumoniae* MTCC109 *Pseudomonas aeruginosa* MTCC1688 *Proteus mirabilis* MTCC1429 *Salmonella typhi* MTCC733 *Salmonella paratyphi* A MTCC735 *Salmonella paratyphi* B *Salmonella typhimurium* MTCC98 *Shigella boydii* *Shigella flexneri* MTCC1457 *Shigella sonnei* MTCC 2957 *Staphylococcus aureus* MTCC737 *Streptococcus faecalis* MTCC459	(Kumaraswamy et al., 2008)
2.	Antioxidant	Leaves	Petroleum ether, chloroform, methanol, and water	-	(Kumaraswamy et al., 2008; Kumaraswamy and Satish, 2008)
		Bark, leaves, and roots	Water, methanol, hexane, petroleum ether, ethanol, dichloromethane, and chloroform	-	(Wani et al., 2018a)
3.	Antidiabetic	Stem wood	Ethanol	Male Sprague Dawley rats	(Ahmad et al., 2008)
3.	Anti-inflammatory	Bark	Ethanol	-	(Kumaraswamy and Satish, 2008; Wani et al., 2018a)

6. Management and Preservation

B. utilis is a versatile plant species that has been used ever since ancient times. The ethno-conservationists have traditionally linked this species to religion in order to preserve and control its over-exploitation. In general, orthodox herbalists and Hindu missionaries gather essential parts of plants in a specific period and prohibit those that do not obey the ritual rules of their collection. Additionally, they recite some mantras for religious or therapeutic purposes while gathering different parts of this species. They do not reveal the therapeutic properties of this tree species. When teaching knowledge to their disciple they are taught not to abuse knowledge and race. This leads to respect for the species, which helps in its preservation (Chandra, 2010).

The current population of this species is under pressure due to extensive collection from designated areas. For example, in the Gangotri region of Uttarakhand, a large number of pilgrims while obtaining Ganges water also debark *B. utilis* for collection of its papery bark. The recurrent over-exploitation of *B. utilis* for fuel wood and other purposes has led to its declining population. Large-scale deforestation from the Himalayan region and excessive exploitation of *B. utilis* trees have been reported, leading to the depletion of its native plantations and habitat loss (Crosby and Worster, 1998).

Conclusion

In this chapter we have highlighted the botanical description, active phytoconstituents, therapeutic uses, and pharmacological activities of *B. utilis*. Traditionally, it has been used for many ethnobotanical purposes. Being an indispensable source of medicine, the conservation of this tree species needs to be governed for its sustained availability and protection of the delicate Himalayan ecosystem. Subsequent studies and research can be done on the plant for its numerous pharmacological activities.

Chapter 2

Corylus avellana L.: Traditional Uses, Phytochemistry, and Pharmacology

Abstract

Corylus avellana L., commonly known as Hazelnut, is one of the most common tree nut species across the globe. The growing interest in hazelnut is due to the importance of its different plant parts such as hazelnut fruits, hazelnut hard coats, hazelnut green leafy coverings, as well as hazelnut leaves. A lot of experimental work has also been carried out to describe the qualitative and quantitative chemical profiles of the various parts of *C. avellana*. The aim of the current chapter is to highlight the botanical aspects, traditional uses, phytochemistry and pharmacology of the species *C. avellana*. In addition to this, habitat and ecology as well as nutritional profile of *C. avellana* has also been highlighted in the current chapter.

1. Introduction

The word hazelnut comes from the Anglo-Saxon word haesel (bonnet). Hazelnut belongs to the genus *Corylus*, species *avellana* that belongs to the family Betulaceae. One of the most common tree-nuts in the world is *Corylus avellana* L. The species' main ingredients are kernels which are rich in nutritious foods with a large concentration of healthy lipids (Alasalvar and Bolling, 2015). The kernels are used by the confectionery sector and consumed fresh (with skin) or ideally roasted (without skin). The kernel is present in a hard shell which in turn is protected by a hazelnut husk and green leaf cover, and also the leaves of the hazelnut tree, are by-products of the roasting, cracking, shelling/hulling and harvesting operations. A broad variety of complex metabolites, recognized as "phytochemicals," that have considerable biological activity, are produced by food plant products. An increasing concern in hazelnut by-products has reportedly been presented in connection with the ongoing research carried out on hazelnut kernels.

It is a monoecious perennial plant that produces an annual crop of nuts, however, they have a strong tendency to bear in alternate years. The plants begin to bear within three years of planting with larger yield after 5-6 years. The plants are pollinated by wind and has a sporophytic self-incompatibility system governed by single gene with multiple alleles. This family includes other genera which include trees and bushes, such as *Betula*, *Alnus*, *Carpinus*, and *Ostrya*. Some experts consider *Corylus* as a member of family Corylaceae, but a new molecular study shows that *Corylus* belongs to the Betulaceae family (APGII system) (Gallego Palacios, 2015). *C. avellana*'s natural distribution is fixed to the northern hemisphere, although it grows in temperate climates such as Turkey, Spain, and Italy.

In recent years, hazelnut production has expanded to new rising areas, including the southern hemisphere while Turkey and Italy remain the main producing countries (80% of world's crops) (Torello Marinoni et al., 2018). It is absent only in Iceland and some Mediterranean islands (Cyprus, Malta, and the Balearic Islands), and in the far north and southeast of the continent (Boccacci et al., 2013). In the mid-1850s, it was introduced into North America (Hummer, 2000). In India, the main area where the tree is found growing is Himachal Pradesh, a hilly northwestern state between latitude 30°22' and 33°12' N and longitude 75°45' to 79°04' E. In this state the species is located in some areas of the Rampur, Rohru, Kotkhai tehsils, Bahali, Sungri, Badseri, and Jareshi area of Shimla district, Sangla (Chansu), Nichar, and Katgaon of Kinnaur district. There are many local names for the nuts, including 'Thangi', 'Gabeja', and 'Burjai'.

C. avellana is a shrub or small tree that can grow up to 8 meters. The bark has white and broad patches and is grey in appearance. The leaves are deciduous, alternate, rounded, 6-12 cm long, each side having a double serrate margin and hairy (Masullo et al., 2015). The flowering takes place very early in the spring before the leaves and is monoecious in nature i.e., both male and female flowers are present on the same plant. The male flowers are borne in catkins that arise in the axils of basal leaves. A small protective bract subtends each individual male flower and 150-200 male flowers on a single stem form the catkin which are green to yellow in colour.

The catkins develop rapidly from August onwards and remain dormant during the winter. The cylindrical catkins are between 3 and 8 cm long and 5 mm in diameter. Pollen grains are about 18-26 microns in size and are a trizonoporate, each pore having a diameter of about 2 microns. Female flowers are borne in tight clusters at three different locations, they are borne singly at the basal leaves on one year wood, in groups of one to six on the catkin peduncles and on very short spurs on the older wood. Once the catkins start to open, 4-18 styles emerge from the top of the bud. They appear like a small red dot at first and then continue to elongate and reflex. If the female flowers remain unpollinated, the stigmas may remain receptive up to 2 months. The ovary is inferior and bicarpellary. Male and female flowers do not open at the same time, usually female flowers bloom in early March and the male flowers bloom after female flowers.

The pollen sheds within a few days after the catkin open. Pollen usually dehisces when the relative humidity drops. In still air, the dehisced pollen will drop on the top of bracts below and is held there until it is blown away by the wind. Once the female flowers have been pollinated, the exposed parts of the styles wither and turns black. Pollen tube grows to the base of style within 4-7 days. As the weather warms up in late spring, the pollen tube continues to grow into the ovary and fertilizes the egg.

After fertilization, the embryo grows for 5-6 weeks and continues to differentiate for 2 more weeks. When the nut has reached full size in July, the wall of the ovary begins to lignify to form a hard shell. The hazelnut fruits are achenes with a brown pericarp containing a single seed that is enclosed by a laciniate boundary involucre (Kole, 2011; Masullo et al., 2016; Rieger, 2006). A dark brown perisperm covers the kernel, the nut of trade, which is covered by a woody shell (Fanali et al., 2018).

1.1. Taxonomical Description

- **Division:** Angiospermae
- **Class:** Magnoliopsida
- **Order:** Fagales
- **Family:** Betulaceae
- **Genus:** *Corylus*
- **Species:** *avellana*
- **Synonyms of *C. avellana*:** *Corylus avellana* f. *atropurpurea* Petz. & G. Kirchn., *Corylus avellana* f. *aurea* Petz. & G. Kirchn. *Corylus avellana* var. *avellana*, *Corylus avellana* var. *davurica* Lideb, *Corylus avellana* f.*glomerata* W.T Aiton, *Corylus avellana* f.*grandis* (Aiton) Lam., *Corylus avellana* f.*laciniata* Petz. & G. Kirchn., *Corylus avellana* var. *macrotruncus* Zernov, *Corylus avellana* f. *microphylla* Gajic & Korac, *Corylus avellana* f.*peltata*

De Langhe, *Corylus avellana* var. *pendula* Goeschke, *Corylus avellana* var.*purpurea* Loudon, *Corylus avellana* var. *rubra* (Borkh.) Lam., *Corylus avellana* f. *variegata* A.DC., *Corylus avellana* var. *zimmermannii* Hahne.

1.2. Vernacular Names

- **English:** Cobnut; European filbert; European Hazel; Filbert; Giant filbert; Haselnut
- **Spanish:** Avellano; Avellano di Lambert
- **French:** Avelinier; Codrier; Noisetier franc; Noisetier tubuleux
- **Portuguese:** Aveleira
- **German:** Gewöhnliche Hasel; Haselnuss; Hasselstrauch; Lambertnuss; Lambertshasel
- **Indian:** Bhotia badam
- **Italian:** Nocciòlo
- **Dutch:** Hazelaar; Gewoone; Hazelnoot, Gewoone
- **Swedish:** Hassel

2. Habitat and Ecology

In mixed deciduous forests, hazelnut typically grows as an understory plant (Kull and Niinemets, 1993). It can grow in both sun (in full light) and shade conditions due to several adaptations at the level of the leaves (Catoni et al., 2015). It grows best in fertile, nutrient-rich soils, which are slightly acidic or neutral or can thrive in dry calcareous soils as well (Clinovschi, 2005). During the growing season, hazelnut prefers mild climates with relatively high temperatures but it can tolerate extreme temperatures or even frosts (Savill, 2019). In Turkey, an average temperature of 13-16°C and precipitation more than 700mm provides optimal conditions for the cultivation of hazelnuts (Ustaoglu and Karaca, 2014).

3. Nutritional Profile of *C. avellana*

Hazelnuts are commonly used in the production of foodstuffs such as chocolate, sweets, bread, ice cream, and dairy products. Hazelnuts can be included with dishes ranging from cereals and bread to yogurt, soups, and salads, and from main dishes to desserts, ice creams, and bakery. They are known for their use in pies and their preference for chocolate (Fallico et al., 2003; Turan and Akccedil, 2011).

Thanks to their unique profile of minerals, proteins, fats, carbohydrates, vitamins, dietary fiber, and phenolic antioxidants, hazelnuts play an important role in nutrition and wellbeing among the groups of dried nuts (Demirbas, 2007; Simsek and Aykut, 2007; Solar and Stampar, 2011). Hazelnuts' nutritional and sensory properties make them a distinctive raw material that is suitable for foods. Because of their very unique nutritional value, hazelnuts can play a significant role in the human diet and wellbeing.

Celluloses and pectins are present at a rate of 1-3% in hazelnuts (Baysal, 1996). One 100 grams of hazelnuts can provide 600-650 kcal, 2–3.5% of minerals, and vitamins to humans (Köksal et al., 2006; Souci et al., 2000). The hazelnut kernel protein content ranges from 10% to 24%. It has been recorded that by eating 100 g of hazelnut per day, 22% of the daily protein intake in the human diet could be supplied (Pala et al., 1996, Alasalvar et al., 2006a; Balta et al., 2006; Cristofori et al., 2008; Garcia et al., 1994; Gunes et al., 2010; Jakopic et al., 2011; Kornsteiner et al., 2006; OÈzdemir et al., 2001). Hazelnut kernels are an excellent source of fats (50–73 percent) and they contain unsaturated fatty acids such as linoleic, linolenic, oleic, palmitic and stearic acids that are essential for human health (Garcia et al., 1994).

Hazelnut oil lowers blood cholesterol levels and regulates the negative impact of elevated blood pressure (Durak, 1999). Minerals are important for maintaining a healthy function of the nerve and for maintaining healthy and well-developed body systems, bones, and cells. Some foods contain minerals that are required by our bodies. Therefore, in the technique of increasing the nutritious content of food while decreasing energy density and consumption, fruits and vegetables attract more popularity (Demigné et al., 2004). Depending on the daily microelement requirements, 100 grams of Turkish hazelnuts will yield around 50% Fe and Cd, 41% Mo, 32% Zn, 21% Se and Cr, 5% B, and 1% Ni of the suggested daily amount (Turan and Akccedil, 2011).

Besides their rich mineral content, the kernels of hazelnut are among the most important sources of essential vitamins, for instance, vitamins B1, B6, niacin, and α-tocopherol. Proof that the active vitamin E-form, α-tocopherol, is gradually contributing to reducing the risk of certain chronic conditions, including heart disease (Iannuzzi et al., 2002), type 2 diabetes (Dhein et al., 2003), hypertension (Boshtam et al., 2002), cancer (Venkateswaran et al.,

2002) and some of the detrimental consequences of ageing may be combated (Grodstein et al., 2003). α-tocopherol can protect against cognitive decline and Alzheimer's disease (Martin, 2003).

4. Uses in Traditional Medicine

Persian medicine, a traditional medicine founded thousands of years ago on Iran's plateau, paid particular interest to hazelnut offering remarkable documented medicinal experiences in cognitive disorders like memory loss and dementia centered that focus on its beliefs about etiology and care. Some texts on Persian medicine emphasized that hazelnuts and almonds strengthen the brain tissue and minimize brain atrophy (Gorji et al., 2018).

In traditional Iranian medicine, *C. avellana*'s leaves, referred to as fandogh, is also widely taken as an efficient liver tonic in the form of infusion (Akbarzadeh et al., 2015); nowadays, owing to its astringency and vasoprotective and anti-edema properties, it is also used in folk medicine for the treatment of haemorrhoids, varicose veins, and phlebitis. Galenic hazelnut leaf extracts have also been used as a treatment for ulcers and oropharyngeal infections. Minor antidysenteric, antifungal, and cicatrizant properties have also been identified (Amaral et al., 2010). In traditional Swedish medicine, hazelnut leaves and bark are used to relieve pain (Amaral et al., 2010).

In Armenia, the decoction of the leaves has also been used in the treatment of liver and gall bladder disorders (Gabrielyan et al., 2004; Isotova et al., 2010). The decoction of the leaves and bark of the trees minimizes blood cholesterol levels, and their infusion is used for varicose vein disease (Gammerman and Grom, 1976; Grossheim, 1952; Gubanov et al., 1976; Tsaturyan and Gevorgyan, 2014; Turova and Sapozhnikova, 1974; Zolotnitskaya, 1965). Vitamins A, B1, C are present in fruits (Grossheim, 1952; Tsaturyan and Gevorgyan, 2007; Zolotnitskaya,1965). *C. avellana* is used for digestive diseases in Azerbaijan, hazelnut shells and dried acorncups are crushed and consumed for diarrhea. Kernel oil is used especially against Ascaris, as anthelmintics (Damirov et al., 1988).

Fried and crushed nut kernels, as well as fresh, are combined and eaten with honey. An emulsion of water is taken internally as enveloping and emollient, with crushed nut kernels, in urolithiasis. With decoction and infusion of the bark, malaria and fever are treated. For general fatigue and low libido, the fried and crushed nut kernels blended with honey are often taken internally. The shells in the decoction of water are called locally "findigchayi" and used for dizziness and against haemorrhoids. Tea from the leaves is seen as reinforcing and stimulating (Damirov et al., 1988). In Georgia, bark, stems, leaves, roots, and oil of all species are used as an antidiarrheal, anti-microbial, anti-inflammatory therapy.

Traditional medicine uses "hazelnut milk" to support people with lung problems, gallstones, and those affected by illness. The honey-mixed fruit is used for anaemia and rheumatism (Kopaliani, 2013). In Akhmeta a cure is made of hazelnut and mistletoe for haemorrhoids and rectum diseases. The leaves act as a cure for gangrene. The fruits are also used for cough treatment (Batsashvili et al., 2017; Bussmann et al.,., 2017a, 2017b, 2016, 2014).

5. Phytochemical Profile

C. avellana has been researched extensively for its phytochemical constituents. The species have been found to be an important source of Phytochemicals, which have antioxidant, anti-inflammatory, anti-cancer, anti-mutagenic, or anti-proliferative properties, and all these compounds play a crucial role and affect human wellbeing. Phenolic compounds, which are researched thoroughly because of their antioxidant properties, are one of the most significant classes of phytochemicals and can be found in different parts of the plant such as skin, shell, leaves, involucrum, and seeds.

The content of phenolic acid and type varies according to the tissues, the techniques of extraction, and the solvents used. The phenolic compounds such as gallic acid, caffeic acid, p-coumaric acid, ferulic acid, and sinapic acid are the most important compounds extracted from the plant. Moreover, the different tissues of *C. avellana* trees also contain flavonoids such as anthocyanidine, flavan-3-ol, proanthocyanidine, and pytoestrogens. These compounds are important mainly to their role in decreasing the risk of cardiovascular disease, cancer, and stroke due to their antioxidant properties (Alasalvar and Shahidi, 2008). The summary of phytochemistry work is as shown in Table 3.

6. Pharmacological Properties

The pharmacological benefits of *C. avellana* are primarily due to bioactive phytochemicals produced in the tissues as primary and secondary metabolites. A detailed list is given in Table 4.

Table 3. Classes and names of compounds isolated from the various parts of *C. avellana*

Sr. no.	Compound	Part of plant	References
1.	3-caffeoylquinic acid, 5-caffeoylquinic acid, caffeoyltartaric acid, and p-coumaroyltartaric acid	Leaves	(Amaral et al., 2005; Oliveira et al., 2007)(Amaral et al., 2010)
2.	Myricetin 3-rhamnoside, quercetin 3-glycoside, quercetin 3-rhamnoside, and kaempferol 3-rhamnoside	Leaves	(Amaral et al., 2010, 2005; Jakopic et al., 2011; Masullo et al., 2016; Oliveira et al., 2007; Riethmüller et al., 2013; Yuan et al., 2018)
	Gallic, 2,4-dihydroxybenzoic, 3,4-dihydroxyibenzoic, 3-hydroxyicinnamic, 2-hydroxybenzoic, syringic, ferulic acid, 3-hydroxy-4-methoxycinnamic, vanillic acid, and caffeic acid.	Kernels	(Prosperini et al., 2009)
	4-hydroxybenzoic,	Kernels	(Prosperini et al., 2009; Zeppa and gerBI, 2010)
	Gallic acid and sinapic acid	Kernels	(Zeppa and geRBI, 2010; Prosperini et al., 2009)(Alasalvar et al., 2006b)
	Caffeic acid, p-coumaric acid, ferulic acid, and sinapic acid	Kernel and green leafy cover	(Alasalvar et al., 2006b)
3.	Gallic acid, caffeic acid, p-coumaric acid, ferulic acid, and sinapic acid	Kernels and skins, hard shells, green leafy covers, and tree leaves	(Shahidi et al., 2007)
4.	p-hydroxybenzoic acid	Kernels	(Slatnar et al., 2014)
5.	Quercetin	Skins	(Del Rio et al., 2011; Ivanović et al., 2020)
6.	Quercetin 3-O-rutinoside, isorhamnetin rutinoside, myricetin rhamnoside, quercetin 3-O-rhamnoside, kaempferol rhamnoside, and phloretin 2'-O-glucoside, myricetin	Skins	(Del Rio et al., 2011)
7.	3-Caffeoylquinic acid, 4-caffeoylquinic acid, p-coumaric acid, myricetin 3-hexoside, and quercetin 3-hexoside	Leaves	(Oliveira et al., 2007)
8.	Quercetin-3-O-rhamnoside and myricetin-3-O-rhamnoside	Kernels	(Jakopic et al., 2011)

Sr. no.	Compound	Part of plant	References
9.	Giffonin P, myricetin-3-O-α-L-rhamnopyranoside, quercetin 3-O- β-D-glucopyranoside, ellagic acid, quercetin 3-O-α-L-rhamnopyranoside kaempferol -3-O- β-D-glucopyranoside, carpinontriol B, giffonin M, kaempferol 3-O-(4''-cis-p-coumaroyl)-α-L-rhamnopyranoside, kaempferol 3-O-(4''-trans-p-coumaroyl)-α-L-rhamnopyranoside, and giffonin Q	Fresh and roasted hazelnut	(Cerulli et al., 2018b)
10.	Gallic, chlorogenic, sinapic, p-coumaric, ferulic acid, ellagic acid, epicatechin, and rutin	Whole nuts	(Solar et al., 2008)
11.	Quercetin 3-O-β-D-galactopyranosyl-(1 → 2)- β -D-glucopyranoside, kaempferol 3-O- β -D-glucopyranosyl-(1 → 2)- β -D-glucopyranoside, quercetin 3-O- β -D-glucopyranoside, quercetin 3-O-α-L-rhamnopyranoside, giffonin I, kaempferol 3-O- α-L-rhamnopyranoside, kaempferol 3-O-(4''-cis-p-coumaroyl)- α-L-rhamnopyranoside, alnusone, giffonin Q, giffonin R, and giffonin S	Flower	(Masullo et al., 2017)
12.	Quercetin 3-O-(2''-O-β-D-glucopyranosyl)-β-D-galactopyranoside,	Pollen	(Strack et al., 1984)
13.	Giffonins T and U	Green leafy covers	(Cerulli et al., 2017)
14.	Protocatechuic acid, syringic acid, and o-coumaric acid.	Kernels	(Zeppa and Gerbi, 2010)
15.	β-Sitosterol, protocatechoic, shikimic acid, p-hydroxybenzoic acid, Isovanillic acid, vanillin, hydroxybenzaldehyde, veratroylglycol, and 3-hydroxy-1-(4-hydroxy-3-methoxyphenyl)propan-1-one (ω-hydroxypropioguaiacone	Shells	(Shataer et al., 2020)
16.	Protocatechuic acid, myricetin 3-O-rhamnoside, quercetin-3-O-rhamnoside, and phloretin-2'-O-glucoside	Kernels (with skin)	(Ciarmiello et al., 2014)
17.	Gallic acid, protocatechuic acid, salicylic acid, syringic acid, vanillic acid, 4-hydroxybenzoic acid, caffeic acid, ferulic acid, o-coumaric acid, and sinapic acid	Natural, roasted, and roasted hazelnut skin	(Pelvan et al., 2018)
18.	Gallic acid	Shells and kernels	(Solar et al., 2008; Yuan et al., 2018)/(Zeppa and Gerbi, 2010)
19.	Protocatechuic acid	Shells	(Solar et al., 2008)
20.	m-hydroxycinnamic acid	Kernels	(Pelvan et al., 2012)
21	Rosmarinic acid	Leaves	(Riethmüller et al., 2013)
22.	Quercetin	Shells	(Yuan et al., 2018)

Table 3. (Continued)

Sr. no.	Compound	Part of plant	References
23.	Quercetin 3-rhamnoside	Kernels, shells, and flower	(Jakopic et al., 2011; Masullo et al., 2016; Yuan et al., 2018)
24.	Epicatechin	Shells and kernels	(Rothwell et al., 2013; Yang et al., 2018)
25.	Catechin	Shells and kernels	(Slatnar et al., 2014; Yang et al., 2018)
26.	Epigallocatechin	Kernels	(Rothwell et al., 2013)
2.	Epicatechin gallate	Shells kernels	(Slatnar et al., 2014; Yang et al., 2018)
28.	Dihydroconiferyl alcohol,	Shells	(Esposito et al., 2017, Shataer et al., 2020)
29.	Vanillic acid	Shells	(Pérez-Armada et al., 2019; Shataer et al., 2020)
30.	Taxifolin, prodelphinidin dimer, procyanidin dimers, and procyanidin trimer	Shells	(Yuan et al., 2018)
31.	Ellagic acid hexoside isomer, ellagic acid pentoside isomer, flavogallonic acid dilactone isomer, bis (hexahydroxydiphenoyl)-glucose (HHDPglucose) isomer, and valoneic acid dilactone/sanguisorbic acid dilactone	Natural hazelnut, roasted hazelnut, and roasted hazelnut skin	(Pelvan et al., 2018)
32.	2,3-dihydroxy-1-(4-hydroxy-3-methoxyphenyl)-propan-1-one, 1-(4-hydroxy-3-methoxy)-1,2,3-propanetriol, threo-1,2-bis(4-hydroxy-3-methoxyphenyl)-1,3-propandiol, giffonin P, erythro-(7S,8R)-guaiacylglycerolβ-O-4'-dihydroconiferyl alcohol, ficusal, erythro-(7R,8S)-guaiacylglycerol-β-O-4'-dihydroconiferyl alcohol, ent-cedrusin, erythro-(7S,8R)-guaiacylglycerol-β-coniferyl aldehyde ether, carpinontriol B, ceplignan, dihydrodehydrodiconiferyl alcohol, giffonin V, balanophonin, kaempferol 3-O-(4'''-cis-p-coumaroyl)-α-L-rhamnopyranoside, and kaempferol 3-O-(4''-trans-p-coumaroyl)-α-Lrhamnopyranoside.	Shells	(Masullo et al., 2017)
33.	Paclitaxel, 10-deacetylbaccatin III, baccatin III, 10-deacetyl7-xylosylcephalomannine, 10-deacetyl-7-xylosylpaclitaxel; 10-deacetyl7-xylosylpaclitaxel C, 10-deacetylpaclitaxel, 7-xylosylpaclitaxel,	Shell	(Ottaggio et al., 2008)

Sr. no.	Compound	Part of plant	References
	cephalomannine, 10-deacetyl-7-epipaclitaxel, paclitaxel C, 7-epipaclitaxel, taxinine M	Leaves and shells	(Ottaggio et al., 2008)
34.	Taxanes	Flowers, and bark	(Jakopic et al., 2011; Masullo et al., 2016)
35.	Kaempferol 3-rhamnoside, quercetin 3-glucoside	Kernels, and bark	(Jakopic et al., 2011; Riethmüller et al., 2013)
36.	Myricetin 3-rhamnoside	Flowers	(Masullo et al., 2016)
37.	Quercetin 3-galactosyl-(1 → 2)-glucoside, kaempferol 3-glucosyl-(1 → 2)-glucoside	Shells and flowers	(Masullo et al., 2017, 2016)
38.	Kaempferol 3-(cis-p-coumaroyl)-rhamnoside, kaempferol 3-(trans-p-coumaroyl)-rhamnoside	Kernels	(Jakopic et al., 2011)
39.	Phloretin-2'-O-glucoside	Kernels	(Rothwell et al., 2013)
40.	Gallocatechin gallate	Leaves	(Riethmüller et al., 2013)
41.	Hirsutenone	Shells	(Masullo et al., 2017)
42.	Carpinontriol B, giffonin V, erythro-(7S,8R)-guaiacylglycerol-β-O-4'-dihydroconiferyl alcohol, erythro-(7S,8R)-guaiacylglycerol-β-coniferyl aldehyde ether, erythro-(7R,8S)-guaiacylglycerol-β-O-4'-dihydroconiferyl alcohol, ficusal, dihydrodehydrodiconiferyl alcohol, and balanophonin	Natural hazelnut, roasted hazelnut, and roasted hazelnut skin	(Pelvan et al., 2018)
43.	(−)-epicatechin-3-O-gallate, eriodictyol, isorhammetin-3-O-rutinoside, and kaempferol-3-O-glucoside	Shells	(Yuan et al., 2018)
44.	Coumaroylquinic, pentose ester of coumaric acid, and Hexose ester of syringic acid	Shells	(Esposito et al., 2017)
45.	Lawsonicin, cedrusin, ficusal, veratric acid, gallic acid, methyl gallate, C-veratroylglycol, and β-hydroxypropiovanillone, and carpinontriol B	Shells	(Pérez-Armada et al., 2019)
46.	Gallic acid, 4-hydroxybenzoic acid, syringic acid, p-coumaric acid, guaiacol, and ellagic acid	Hazelnut Skins	(Ivanović et al., 2020)
47.	Galic acid, protocatechuic acid, p-hydroxybenzoic acid, ferulic acid, catechin epicatechin, rutin, resveratrol, and naringenin		

Table 4. An overview of the modern pharmacological assessments of *C. avellana*

S. No	Activity	Parts used	Extract	Organism used/Cell lines	Reference
1.	Antimicrobial	Hazelnut skins	Methanol	*Lactobacillus plantarum* P17630 and *Lactobacillus crispatus* P17631	(Montella et al., 2013)
		Green leafy covers	Methanol	*Bacillus cereus, Bacillus cereus, Escherichia coli, Pseudomonas aeruginosa,* and *Staphylococcus aureus*	(Cerulli et al., 2017)
		Fresh Hazelnuts (Tannins)	Acetonic	*Listeria monocytogenes, S. aureus, E. coli, Brochothrix thermosphacta, P. fragi, Salmonella* Typhimurium, and *L. plantarum*	(Amarowicz et al., 2008)
		Oils from the nuts	Ethanol:hexane (1:1)	*E. coli, P. aeruginosa, Proteus mirabilis, Klebsiella pneumoniae, Acinetobacter baumannii, S.aureus, Enterococcus faecalis, Candida albicans,* and *C. parapsilosis*	(ORHAN et al., 2011)
		Leaves	Aqueous	*B. cereus, B. subtilis, S. aureus, P. aeruginosa, E. coli, Klebsiella pneumonia, Candida albicans,* and *Cryptococcus neoformans*	(Oliveira et al., 2007)
7.	Antioxidant	Kernel and hazelnut green leafy cover	Ethanol/water mixture and acetone/water mixture	-	(Alasalvar et al., 2006b)
		Kernels	Aqueous	-	(Delgado et al., 2010)
		Oils from the nuts	Ethanol:hexane (1: 1)	-	(ORHAN et al., 2011)
		Green leafy covers	Methanol	-	(Cerulli et al., 2017)
		Whole raw and roasted hazelnuts	Hexanes	-	(Li and Parry, 2011)
		Natural hazelnut, roasted hazelnut, and roasted hazelnut skin	Acetone/water/acetic acid mixture (70:29.5:0.5)	-	(Pelvan et al., 2018)
		Kernels and hazelnut skins	Ethanol	-	(Shahidi et al., 2007)

S. No	Activity	Parts used	Extract	Organism used/Cell lines	Reference
		hazelnut hard shells, hazelnut green leafy covers, and hazelnut tree leaves			
		Kernels	Water, methanol, and acetone	-	(Delgado et al., 2010)
		Hazelnut skins	Hexane and chloroform/methanol	-	(Ivanović et al., 2020)
		Hazelnut skins	Methanol	-	(Montella et al., 2013)
		Shells	Methanol	-	(Esposito et al., 2017)
		Kernels	Ethanol, methanol, and acetone	-	(Zeppa and Gerbi, 2010)
		Kernels	Methanol 0.1% HCl, ethanol 0.1% HCl, acetone 0.1% HCl, methanol/water (8:2, v/v), ethanol/water (8:2, v/v), acetone/wate (8:2, v/v)	-	(Fanali et al., 2018)
		Leaves	Aqueous	-	(Oliveira et al., 2007)
	Cytotoxic	Shells	Methanol	Human malignant melanoma (A375), and Human cervical cancer (HeLa)	(Esposito et al., 2017)
		Whole raw and roasted hazelnuts	Hexanes	HT-29 human colorectal adenocarcinoma	(Li and Parry, 2011)
		Leaves	Ethanol and water	THP-1 cells	(Cerulli et al., 2018a)
		Stems and leaves	Methanol	HeLa, human cervical cancer cells; HepG2, liver hepatocellular cells; MCF-7, human breast adenocarcinoma cell	(Cerulli et al., 2018a)

Conclusion

This chapter summarizes the results from the current studies of *C. avellana* which is one of the important medicinal sources for herbal drugs. The plant has been used extensively as a source of traditional medicine for the treatment of various disorders like memory loss and dementia and have found to strengthen the brain tissue and minimize brain atrophy. Extracts derived from the different sections of *C. avellana* along with pure compounds have demonstrated many biological functions, including antioxidant, antiproliferative, antimicrobial, and neuroprotective effects.

Chapter 3

Ficus carica L.: Traditional Uses, Phytochemistry, and Pharmacology

Abstract

Ficus carica is a blessed and essential medicinal plant that has been largely used in many countries as a traditional medicine in the treatment and prevention of many complications. The most common bioactive compounds in this plant are flavonoids and various plant extracts have been found to possess many biological activities. Fig leaves contain polyphenols that are potentially beneficial for human health with antioxidant and radical scavenging effects. Traditionally, fig leaves have been used for treating diabetes and liver disorders. The current review highlights the botanical description, nutritional value, traditional uses, phytochemistry and pharmalogical aspects of *F. carica*.

1. Introduction

The genus *Ficus* (Moraceae) constitutes one of the largest genera of angiosperms with more than 800 species of trees, epiphytes, and shrubs which are distributed in the tropical and sub-tropical regions of the world (Singh et al., 2011). The genus is divided into six subgenera according to its preliminary morphology. *Ficus* comprises 23 hemiepiphyte and lithophyte species that possess aerial and creeping root systems (Rønsted et al., 2008). It is one of the 40 genera of the Moraceae family.

F. carica has been grown for a long time for its edible fruit in different locations nationwide. The major producers of edible figs are Turkey, Egypt, Morocco, Spain, Greece, California, Italy, Brazil and other places with usually mild winters and long, dry summers (Rahmani and Aldebasi 2017). *F. carica* is one of the oldest fruit cultivated species and is, currently an important crop worldwide, and is thought to have been originated from the Middle East Zohary et al., 2012). The common fig still grows wild in the Mediterranean basin nowadays. The Asian-Australian zone, comprising around 500 *Ficus* species, is the richest and most diverse. In comparison, African *Ficus* and Neotropics are less common, with 110 and 130 species, respectively. Around half of the species of *Ficus* are monoecious and the remainder are dioecious in nature (Berg, 2003). Several species of *Ficus* consist of various varieties, possess considerable genetic diversity, and extraordinary commercially relevant pharmacological activities (Woodland, 1991).

The most common member of the genus *Ficus* is *F. carica* Linn and the plant is native to the Sub-Himalayan region, Bengal and central India, and has also been grown extensively worldwide. It is a temperate species, native to southwest Asia and the Mediterranean region (from Afghanistan to Portugal) and has been extensively grown for its fruits (nutritional value), also known as figs, since prehistoric times. Throughout the tropics, figs have great cultural significance, both as objects of worship and for their many practical uses.

The dried fruits of *F. carica* have significant source of vitamins, minerals, carbohydrates, sugars, organic acids, and phenolic compounds (Shahinuzzaman et al., 2020; Mawa et al., 2013). There are also high concentrations of fibre and polyphenols in the fresh and dried figs (Vinson, 1999). Figs are an outstanding source of phenolic compounds such as proanthocyanidins, while red wine and tea contain phenols lower than those in figs, which are two strong sources of phenolic compounds (Vinson et al., 1999).

This plant also draws interest to the biological activities of researchers worldwide. In traditional systems of medicine such as Ayurveda, Unani, and Siddha, the therapeutic utilities of *F. carica* have been established (Prasad et al., 2006). It has been used to treat endocrine system disorders (diabetes), respiratory system disorders (liver diseases, asthma, and cough), gastrointestinal tract (ulcer and vomiting), reproductive system disorders (menstrual pain), and infectious diseases (skin disease, scabies, and gonorrhoea) (Badgujar et al., 2014).

The structures of fruit and reproduction of species of the genus *Ficus* are an exception. It can only be pollinated by their related agaonid wasps, and in exchange the wasps can only produce eggs within the related fruit. In order for the species of *F. carica* to be efficiently pollinated and replicated, its corresponding pollinator wasp should be present. Alternatively, *Ficus carica* and its related species should be present in order for their efficient reproduction by agaonid wasps (Janzen, 1979). The *Ficus carica* pollinator wasp is *Blastophaga psenes* (L.) (Wagner et al., 1990).

Ficus carica L.

1.1. Taxonomical Description

- **Kingdom:** Plantae
- **Subkingdom:** Viridiplantae
- **Infrakingdom**: Streptophyta
- **Superdivision:** Embryophyta
- **Division:** Tracheophyta
- **Subdivision:** Spermatophytina
- **Class:** Magnoliopsida
- **Superorder:** Rosanae
- **Order:** Rosales
- **Family:** Moraceae
- **Genus:** *Ficus*
- **Species:** *Ficus carica*
- **Synonyms:** *Ficus carica* var. *afghanica* Popov, *Ficus carica* var. *caprificus* Risso, *Ficus carica* var. *domestica* Czern. & Rav., *Ficus carica* var. *globosa* Hausskn., *Ficus carica* var. *johannis* (Boiss.) Hausskn., *Ficus carica* var. *longipes* Bornm. ex Parsa, *Ficus carica* var. *riparium* Hausskn., *Ficus carica* var. *rupestris* Hausskn.

1.2. Vernacular Names

F. carica is known by more than 135 names in different countries or parts of the world.

- **Arabic:** Anjir, Teen, Teen barchomi, Ten
- **Brazil:** Figo, Figueira, Figueira-Da-Europa
- **Burmese:** Thaphan, Thinbaw, Thapan
- **Chinese:** Mo Fa Guo, Wu Hua Guo
- **Cook Islanda:** Suke
- **Croatian:** Smokva, Smokvencia, Smokvina
- **Czech:** Smokvon
- **Danish:** Almindelig Figen, Figen
- **Dutch:** Echte Vijeboom, Gewone Vijgeboom, Vijg
- **Eastern Indian:** Doomoor, Angir, Dumur, Udumbara
- **Eastonian:** Harilik, Viigipuu

- **English:** Common fig tree, Fig
- **Eucadorian:** Higo
- **Finnish:** Viikuna
- **French:** Caprifiguier, Carique, Figue, Figuier
- **German:** Echte Feige, Echter Feigenbaum, Essfeige, Feige, Feigenbaum
- **Hindi:** Anjeer, Anjir, Tin
- **Hungarian:** Fugea
- **Iran:** Anjeer
- **Italian:** Fic, Fico, Fico Comune
- **Japanese:** Ichijiku
- **Korean:** Mu Hwa Gwa, Mu Hwa Gwa Na Mu
- **Macedonian:** Smoka
- **Malaysia:** Anjir
- **Mangarevan:** Pika
- **Marshallese:** Wojke, Piik
- **Nepalese:** Anjiir
- **Northern Indian:** Fagari
- **Norwegian:** Fiken
- **Pakistan:** Faag, Anjeer, Injir, Baghi, Inzar, Anzar, Anjir
- **Palauan:** Uosech
- **Palestinian:** Fig
- **Persian:** Anjir, anjeer
- **Polish:** Figowiec
- **Portuguese:** Figueira-Do-Reino
- **Russian:** Inzir
- **Samoan:** Mati
- **Sanskrit:** Angira, Anjeer, Anjir, Anjira, Phalgu, Rajodumbara, Udumvara
- **Serbian:** Smoka, Smokovnica, Smokva
- **Solvascina:** Figa, Figovec, Figovina
- **Solvencina:** Figonik
- **Southern Indian:** Anjeera, Anjoora, Anjooramu, Anjura, Anjuru, Appira, Cevvatti, Chikappatti, Cimaiyatti, Madipatu, Manchi Medi, Manjimedi, Shima-Atti, Shimayatti, Simaatti, Simayatti, Simeyam, Simmeatti, Tacaiyatti, Tenatti, Tenatti, Teneyatti, Theneyatthi, Utumparam

- **Spanish:** Breva, Higo, Hibuera comun
- **Swedish:** Fikon, fikontrad
- **Western Indian:** Anjir, Anjeer, Angir
- **Urdu:** Poast, Darakht Anjir, Anjir Zard
- **Unani:** Anjir
- **Tongarevan:** Monamona
- **Taumotuan:** Tute
- **Turkey:** Incir, Yemis
- **Vietnamese:** Qua Va, Vo Hoa Qua(Khare, 2007; Nadkarni, 1982)

2. Botanical Description

F. carica, with many spreading branches from a short, rugged trunk, is typically 5-9m tall deciduous tree. The species has smooth, grey or dull white bark, and glabrous or softly haired young twigs. Leaves are up to 12 cm long with grooved petiole, glabrous to tomentose. The lamina is variable in shape and size, broadly ovate to almost orbicular. Hypanthodia axillary solitary or paired, borne on up to 3 cm long peduncles, pyriform to globose, 1.5-2 cm in diameter, subsessile to sessile, subtended by 3 broadly deltoid basal bracts.

Male flowers typically consist of 4 sepals, which are united, lobes are lanceolate, stamens are 4 in number with long filaments and exserted anthers. Female flowers are pedicellate with 4 sepals, lobes are lanceolate-oblong, ovary with lateral style, stigma is entire or bifid. Figs are usually pyriform-obovoid in shape, 2-5 cm in diameter, glabrous or shortly hispid, yellowish to brownish violet.

Seeds can be large, medium, small, or minute and range in number from 30 to 1,600 per fruit. Until pollinated, the edible seeds are numerous and usually hollow. Pollinated seeds provide dried figs with the characteristic nutty favour. The interior portion is a white, inner ring containing a seed mass bound with jellylike flesh (Flora of Pakistan, Flora of China). The plant's latex is milky white and contains ficin, i.e., protein hydrolytic enzyme (Badgujar, 2011). Usually, the root system in the plant is shallow and spreading. The species name *carica* indicates having papaya-like leaves.

3. Traditional Use of *Ficus carica*

F. carica is used as a mild laxative, expectorant, diuretic in Unani medicine, and also as a deobstructive and anti-inflammatory agent in the treatment of liver and spleen diseases (Duke 2002). Leaves, fruits, and roots of *F. carica* are used in various disorders such as gastrointestinal (colic, indigestion, loss of appetite and diarrhoea), respiratory (sore throats, cough and bronchial problems), inflammatory and cardiovascular disorders in the native medicinal method (Burkill, 1935; Penelope, 1997). For the prevention of leucoderma and ringworm infection, roots are sometimes used as a tonic.

F. carica in combination with *Laurus nobilis* Linn (Lauraceae), another medicinal plant commonly known as Paathri, and with natural foods such as honey and milk (Idolo et al., 2010; Ratna et al., 2011). Latex is used as an expectorant, anthelmentic, diuretic, and anaemia. Furthermore, the fig is so commonly used both fresh and dried in Mediterranean nations that it is called "the food of the poor man". As it is free of fat and cholesterol and contains a high number of amino acids (Solomon et al., 2006; Veberic et al., 2008; Slatnar et al. 2011).

The fruit juice of *F. carica* combined with honey is used for haemorrhages (Solomon et al., 2006). It is used as an aid in liver and spleen diseases. The dry fruit of *F. carica* is a supplement food for diabetics. It is sold on the market as sweet because of its high sugar content (Veberic et al., 2008). Fruit paste is applied to swellings, tumours, and inflammation for treating pain. Seeds are utilized as edible oil and lubricant (Kirtikar & Basu, 1995). Table 5 shows conventional uses of *F. carica*, including Ayurvedic, Unani, and reports of various ethnobotanical surveys.

4. Nutritional Value

F. carica is commonly grown for its edible fruits. Because of its edible fruit, it is commonly cultivated. These are sugary and juicy; a fully ripe product is an incredible fruit that melts nearly completely in the mouth. For later usage, the fruit is also dried and this dried fruit is an important trade commodity.

Figs are one of the greatest sources of calcium and fiber. Dried figs are richer in fiber, copper, manganese, magnesium, potassium, calcium, and vitamin K, in accordance with human requirements, according to USDA data for the Mission variety.

Table 5. Traditional uses of *Ficus carica*

S. no.	Uses	Part/Preparation used	Locality	Reference
1.	Abdominal pain	Decoction with dried fruits and unpeeled almond	Abruzzo, Italy	Idolo et al. (2010)
	Anthelmentic	Fruits are used a tonic	Gilgit, Pakistan	Khan and Khatoon (2007)
2.		Latex	Peshawar, Pakistan	Zabihullah et al. (2006)
3.	Antiseptic for urinary tract	Decoction made with 0.5 l water, 5dried fruits, 4 Laurus nobilisleaves and a peeled apple	Abruzzo, Italy	Idolo et al. (2010)
4.	Anticancer	Fruits	-	Rubnov et al., 2001;Pérez et al., 1999
4.	Anorexia	Bark, leaves and latex	Pakistan	Ikram et al. (2014)
5.	Anemia	Fruits	Khouzestan, Iran	Ramazani et al. (2010)
6.	Antidiarrheal	Fruits	-	Duke et al., 2002, Werbach 1993
7.	Anti-inflammatory	Fruits	-	
8.	Anthelmintic	Leaves		Saeed and Sabir, 2002
9.	Antioxidant	Fruits	-	Penelope, 1997
10.	Antiplatelet, inflammatory, and gut motility	Fruits	Pakistan	Gilani et al., 2008
11.	Antispasmodic	Fruits		Duke et al., 2002, Werbach 1993, Gilani et al., 2008
12.	Ascites	Fruit	Khouzestan, Iran	Ramazani et al. (2010)
13.	Bee sting	Latex soothes the bee sting by simply rubbing on the skin	Buner, Pakistan	Badgujar (2011)
14.	Blood deficiency	Leaves	South-eastern Nigeria	Nebedum et al. (2010)
15.	Boils	Fruit extraction	Istanbul, Turkey	Genç and Özhatay 2006
16.	Bone treatment	Bark	India	Kirtikar and Basu (1995)
17.	Bronchitis	Aqueous infusion of fresh leaf tender is taken orally as a drink	Ecuador	Tene et al. (2007)

Table 5. (Continued)

S. no.	Uses	Part/Preparation used	Locality	Reference
18.	Burn and Emollient	Fruit and latex	Nablus, Palestine	Jaradat (2005)
19.	Cardiac troubles	Fruits are used a tonic	Gilgit, Pakistan	Khan and Khatoon (2007), Duke et al., 2002, Werbach 1993,
20.	Colic treatment	Fruit, root, and leaf	-	Burkill, 1935, Penelope, 1997
21.	Corns	Latex	Istanbul, Turkey	Genç and Özhatay 2006
22.	Constipation	Juice extracted from fruit is taken orally	Jodhpur, India	Prajapati et al., (2007)
23.	Cough	Decoction of fruit with honey	Abruzzo, Italy	Idolo et al. (2010)
		Leaves	Malaysia	Mohamad et al., 2011
		Fruit and latex	Nablus, Palestine	Jaradat (2005)
		Decoction of boiled fruits is taken orally	Northern and central Oman	Ghazanfar and Al-Abahi (1993)
24.	Diabetes	Decoction of leaves	Islamabad, Pakistan	Khan et al. (2011)
25.	Drink (Tea)	Dry fruit powder is used in tea recipes, as nutritional one	Turkey	Uysal et al. (2010)
26.	Earache	Leaf juice with honey	Nepal	Kunwar and Bussmann (2006)
27.	Eye vision problem	Powder of dry fruits and sugar is taken orally with water twice a day	Abbottabad, Pakistan	Abbasi et al. (2010)
28.	Expectorant	Fruits	Northern Pakistan	Afzal et al., (2009)
29.	Fever	Dried fruits	Bangladesh	Khanom et al. (2000)
30.	Food	Fruits	Shangla, Pakistan	Ibrar et al. (2007)
		Dried fruit to sweeten decoction	Abruzzo, Italy	Idolo et al. (2010)
		Leaves are fodder for goats	Buner, Pakistan	Badgujar (2011)
		The unripe fruit and young growth are cooked and eaten as vegetable	Rawalpindi, Pakistan	Husain et al. (2008)
		Fuel Wood	Buner, Pakistan	Badgujar (2011)
31.	Hemorrhoids	Leaves	Istanbul, Turkey	Genç and Özhatay 2006
32.	Hepatitis	Decoction of fruit	Istanbul, Turkey	

S. no.	Uses	Part/Preparation used	Locality	Reference
33.	Indigestion	Fruits, root, and leaves	-	Burkill, 1935, Penelope, 1997
34.	Inflammation	Bark	Khouzestan, Iran	Ramazani et al. (2010)
35.	Intestinal pain	Bark	Pakistan	Marwat et al. (2009)
36.	Irritant potential	Leaves		Saeed and Sabir, 2002
37.	Jaundice	20 ml of leaf juice mixed with a cup of goat milk is administered early in the morning one a day for 3 d	Andhra Pradesh, India	Manjula et al. (2011)
38.	Kidney stone	Fruit and latex	Nablus, Palestine	Jaradat (2005)
		Bark and leaves	Pakistan	Marwat et al. (2009)
		Fruits	Northern Pakistan	Afzal et al. (2009)
39.	Laxative	Leaves, latex, Fig	Buner, Pakistan	Sher et al. (2011)
		Fruit juice is taken orally	Jodhpur, India	Prajapati et al. (2007)
40.	Leucoderma	Root	Maharashtra, India	Kalaskar et al. (2010)
41.	Liver diseases	Fruits	Northern Pakistan	Afzal et al. (2009)
42.	Loss of appetite	Fruits, root, and leaf	-	Burkill, 1935; Penelope 1997
43.	Menstruation pain and Sedative	Aqueous infusion of fresh leaf tender is taken orally as a drink	Eucador	Tene et al. (2007)
44.	Metabolic	Fruits	-	Duke et al., 2002, Werbach, 1993
45.	Mild laxative, expectorant, and diuretic	Fruits	India	Khadabadi et al., 2007
46.	Mouth cavity diseases	Dried fruits	Abbottabad, Pakistan	Qureshi et al. (2008)
47.	Nutritive diet	Fruits	Mediterranean countries	Çalişkan and Polat, 2011.
48.	Paralysis	Dried fruit	Bangladesh	Khanom et al. (2000)
49.	Piles and chronic ulcer	Fruit juice and latex	Rawalpindi, Pakistan	Husain et al. (2008)
50.	Prevention of nutritional anaemia	Leaf	-	Saeed and Sabir, 2002
51.	Regulates blood stream	Decoction made with dried fruits, lemon peel and *Laurus nobilis* leaves	Abruzzo, Italy	Idolo et al. (2010)
52.	Respiratory	Fruits		Duke et al., 2002, Werbach 1993

Table 5. (Continued)

S. no.	Uses	Part/Preparation used	Locality	Reference
53.	Skin disease	Fruit and latex	Nablus, Palestine	Jaradat (2005)
		Stem latex	Gilgit, Pakistan	Khan and Khatoon (2007)
		Fruit and latex	Nablus, Palestine	Jaradat (2005)
54.	Stomach cancer	Leafy latex	Sari, Iran	Hashemi et al., 2011
55.	Tuberculosis	Leaf	Malaysia	Khadabadi et al., 2007
56.	Wart	Milky latex is applied externally	Alasehir, Turkey. Izmir, Turkey. Jodhpur, India. Istanbul, Turkey.	Ugulu (2011); Ugulu et al. (2009); Prajapati et al. (2007); Genç and Özhatay 2006
		Fruit juice	Kashmir, Pakistan	Ishtiaq et al. (2007)
57.	Weakness	Single dry fruit is deep in water for a night. This fruit is consumed at morning for 15 d	Maharashtra, India	Patil et al., (2011)
58.	Wound	Fruit and latex	Nablus, Palestine	Jaradat (2005)

Figs have a mineral content that strongly matches that of human milk. They have many other nutrients in smaller quantities. Compared with apples (7.14 mg/100 g), bananas (3.88 mg/100 g), dates (25.0 mg/100 g), grapes (10.86 mg/100 g), oranges (40.25 mg/100 g), prunes (18.0 mg/100 g), raisins (40.0 mg/100 g), and strawberries (14.01 mg/100 g), figs produce more calcium (132.5 mg/100 g). The average energy intake dramatically decreased following the supplementation of fiber-rich food, according to Pasman et al. (1997), the average energy intake decreased significantly after the supplementation of fiber-rich food.

A recent study has shown that the addition to the dietary food material of a soluble fibre supplement may help in weight loss. Figs and their soluble fibre can also be helpful in reducing weight, since figs have more fibre (12.21 g/100 g) than all the common fruits mentioned above (Vinson, 1999). Owing to their contribution to the taste, colour, and nutritional properties of fruits, phenolics are an essential component of fruit quality. Rutin (28.7 mg/100 g) is found in the highest quantities among the phenolics studied in the fruit of this plant (Veberic et al., 2008), followed by (+) -catechin (4.03 mg/100 g), chlorogenic acid (1.71 mg/100 g), (-)-epicatechin (0.97 mg/100 g), gallic acid (0.38 mg/100 g) and, lastly, syringic acid (0.10 mg/100 g). As a traditional, seasonal fresh fruit, figs are therefore an integral component of the regional diet.

Figs are also an outstanding source of carbohydrates, minerals, lipids, and enzymes (Table 6). It is high-carbohydrate food. In dried figs, ninety-two percent of the carbohydrates are glucose, fructose and sucrose. The remainder consists of dietary fibre, insoluble cellulose in the skin, soluble pectin in fruit. From the fruit of the fig tree, numerous lipid compounds were established. Triacylglycerols, free and esterified sterols, mono-and digalactosyl diglycerides, ceramide oligosides, cerebrosides, esterified sterol glycosides and phosphatidyl glycerols are the major groups. Myristic acid, palmitic acid, stearic acid, oleic acid, linoleic acid and linolenic acid are the most common fatty acids in fig (Marwat et al., 2009). For commercial purposes, several proteolytic enzymes such as diastase, esterase, lipase, catalase and peroxidase (Jeong and Lachance, 2006) are isolated from figs. The latex acquired from stem of fig plant include enzymes such as ficin, proteases, lipodiastases, amylase where as tyrosin, cravin, lipase, are found in skin of the fig (Canal et al., 2000).

Table 6. Nutritional composition of *Ficus carica*

Protein	1.3 g
Fat (total lipids)	0.2 g
Carbohydrates	7.6 g
Calories (energy)	80 k Cal
Moisture	88.1 g
Fiber	2.2 g
Minerals	0.6 g
Thiamine	0.1 mg
Calcium	35 mg
Phosphorous	22 mg
Iron	0.6 mg
Vitamin A	80 IU
Vitamin C	2 mg

5. Phytochemistry

The existence of various bioactive compounds such as phenolic compounds, phytosterols, organic acids, anthocyanin composition, triterpenoids, coumarins, and volatile compounds such as hydrocarbons, aliphatic alcohols and a few other groups of secondary metabolites was revealed by phytochemical studies on *F. carica*. Some of the *F. carica* phytoconstituents are used in the manufacture of sunscreen and colouring agents (Chawla et al., 2012). Table 7 includes a number of compounds isolated from different parts of the plant.

6. Pharmacological Properties

Fig has been reported to exhibit anticancer, antioxidant, anti-diabetic, anti-fungal, anti-mutagenic activity (Gilani et al., 2008). A detailed list is given in Table 8.

Table 7. Phytochemicals present in *Ficus carica* plant

Sr. no.	Compound	Part of plant	References
1.	Quercetin, luteolin, biochanin-A, luteolin-6C-hexose-8C-pentose, apigenin rutinoside, kaempferol rutinoside, quercetin rutinoside, quercetin glucoside, quercetin acetilglucoside, 3-CQA (3-*O*-caffeoylquinic acid), 5-CQA (5-*O*-caffeoylquinic acid), Q-3-Glu (quercetin 3-*O*-glucoside), Q-3-rut (quercetin 3-*O*-rutinoside), ferulic acid, psoralen, bergapten, pyrogallic, phenol, 3–5-dimethoxy, coumaric, phenolptethlin, pinocembrine, chysin, galangin, protocetchol, vinallin, cinnamic, quercetin, pinostrobin, bauerenol, calotropenyl acetate, lupeol acetate, methyl maslinate, oleanolic acid, β-sitosterol, umbelliferone, 3-Methyl-butanal, 2-Methyl-butanal, *(E)*-2-Pentenal, Hexanal, *(E)*-2-Hexenal, 1-Penten-3-ol, 3-Methyl-1-butanol, 2-Methyl-1-butanol, 1-Heptanol, Benzyl alcohol, (E)-2-Nonen-1-ol, Phenylethyl alcohol, 3-Pentanone, Methyl butanoate, Methyl hexanoate, Hexyl acetate, Ethyl benzoate, Methyl salicylate, Limonene, Menthol, α-Cubenene, α-Guaiene, α-Ylangene, Copaene, α-Bourbonene, β-Elemene, α-Gurjunene, β-Caryophyllene, β-Cubebene, Alloaromadendrene, α-Caryophyllene, Υ-Muurolene, Germacrene D, (+)-Ledene, Υ-Elemene, Υ-Cadinene, α-Muurolene, β-Cyclocitral, Υ-Nonalactone, Psoralen, Oxalic, citric, malic, quinic, shikimic and fumaric acids	Leaves	Kang et al., 1995; Vaya and Mahmood 2006; El-Shobaki et al., 2010; Vallejo et al., 2012
2.	Betulol, lupeol, lanosterol, lupeol acetate, β-amyrin, β-sisterol, α-amyrin, myristic acid, pentadecylic acid, palmitic acid, margaric acid, *cis*-10 heptadecenoic acid, stearic acid, oleic acid, elaidic acid, linoleic acid, arachidic acid, heneicosylic acid, behenic acid, tricosylic acid, lignoceric acid, leucine, tryptophan, phenylalanine, lysine, histidine, asparagine, alanine, glutamine, serine, glycine, ornithine, tyrosine, cysteine, pentanal, hexanal, heptanal, benzaldehyde and octanal 7, 1-butanol-3-methyl, 1-butanol-2-methyl, 1-pentanol, 1-hexanol, 1-heptanol, phenylethyl alcohol, phenylpropyl alcohol; 6-methyl-5-hepten-2-one; monopterpenes α-thujene, α-pinene, β-pinene, limonene, terpinolene, eucalyptol, cis-linalool oxide, linalool and	Latex	Oliveira et al., 2010; Rubnov et al., 2011.; Ghanbari et al., 2019

Table 7. (Continued)

Sr. no.	Compound	Part of plant	References
	Epoxylinalool; sesquiterpenes α-guaiene, α-bourbonene, β-caryophyllene, trans- α-bergamotene, α-caryophyllene, Υ-muurolene, germacrene D, cadinene, α-calacorene, 6-O-Acyl-β-D-glucosyl-β-sitosterols, glycosides - 6-O-acyl-β-D-glucosyl-β-sitosterols		
	Cyanidin-3-rhamnoglucoside, cyanidin-3-glucoside, cyanidin-3,5-diglucoside, cyanidin-3-rutinoside, pelargonidin-3-glucoside, cyanidin 3-rutinoside dimmer, (epi)catechin-(4 → 8)-cyanidin 3-glucoside, (epi)catechin-(4 → 8)-cyanidin 3-rutinoside, cyanidin 3,5-diglucoside, (epi)catechin-(4 → 8)-cyanidin 3-rutinoside, (epi)catechin-(4 → 8)- pelargonidin 3-rutinoside, (epi)catechin-(4 → 8)-pelargonidin 3-rutinoside, carboxypyrano-cyanidin 3-rutinoside, cyanidin 3-malonylglycosyl-5-glucoside, Pelargonidin 3-glucoside, pelargonidin 3-rutinoside, peonidin 3-rutinoside, cyanidin 3-malonylglucoside, campesterol, stigmasterol, sitosterol, fucosterol, 3-Methyl-butanal, 2-Methyl-butanal, 2-Methyl-2-butanal, (E)-2-Pentenal, Hexanal, (E)-2-Hexenal, Heptanal, (Z)-2-Heptanal, Benzaldehyde, (E, E)-2,4-Heptadienal, Octanal, (E)-2-Octenal, Nonanal, (E, Z)-2,6-Nonadienal, 1-Penten-3-ol, (Z)-3-Hexen-1-ol, 3-Methyl-1-butanol, Benzyl alcohol, (E)-2-Nonen-1-ol, Phenylethyl alcohol, 6-Methyl-5-hepten-2-one, Methyl hexanoate, Methyl salicylate, Ethyl salicylate, α-Pinene, β-Pinene, Limonene, Eucalyptol, Linalool, Epoxylinalool, Menthol, α-Cubenene, Copaene, β-Caryophyllene, (E)-α-Bergamotene, Υ-Muurolene, Germacrene D, Υ-Cadinene, Parkeol	Fruit	Solomon et al., 2006; Dueñas et al., 2008; Yemiş et al., 2012; Jeong and Lachance 2001; Oliveira et al., 2010

Table 8. An overview of the modern pharmacological assessments of *Ficus carica*

S. no	Activity	Parts used	Extract	Organism used/Cell lines	Reference
1.	Antipyretic	Leaves	Ethanol	Albino rats	Patil and Patil, 2011
2.	Anti-inflammatory activities	Leaves	Methanol	Male Wistar rats	Eteraf-Oskouei et al., 2015
					Allahyari et al., 2014
			Petroleumether, chloroform, and ethanol	Albino rats	Patil and Patil, 2011
			50% ethanol + 50% distilled (v/v)	Male Wistar rats	Ali et al., 2012
		Branches	Ethanol	-	Park et al., 2013
3.	Antioxidant	Leaves	50% ethanol + 50% distilled (v/v)	-	Ali et al., 2012
		Fruit, leaves, and stem bark	Ethanol	-	Mopuri and Islam 2016
		Leaf, peel, and pulp	70% ethanol	-	Ara et al., 2020
		Fruits	Aqueous	-	Yang et al., 2009
		Leaves	Methanol	-	Ahmad et al., 2013
		Fruits	Methanol	-	Harzallah et al., 2016
		Fruits	Aqueous-methanol 80% (v/v)	-	Hoxha et al., 2015
		Fruits	80% methanol, 60% acetone, aqueous, petroleum ether	-	Debib et al., 2014
		Latex	Methanol and ethanol	-	Shahinuzzaman et al., 2020
		Leaves	Methanol	-	Mahmoudi et al., 2016
		Branches	Ethanol	-	Park et al., 2013
4.	Antispasmodic and antiplatelet	Fruits	Aqueous ethanol (80%)	Adult rabbit	Gilani et al., 2008

Table 8. (Continued)

S. no	Activity	Parts used	Extract	Organism used/Cell lines	Reference
5.	Anthelmintic	Leaves	Petroleum ether, chloroform, methanol, and water	*Pheretima posthuma*	Amol et al., 2010
		Fruits	Aqueous	*P. posthuma*	Sawarkar et al., 2011
		Bark	Petroleum ether, ethyl acetate, and methanol	*P. posthuma*	Tambe et al., 2009
		Latex	-	*Syphacia obelata, Aspiculuris tetraptera, Vampirolepis nana.*	de Amorin et al., 1999
6.	Cytotoxic	Latex	Methanol, hexane, ethyl acetate, chloroform	*Cercopithecus aethiops*	Lazreg Aref et al., 2011
			-	Esophagus cancer cell line	Hashemi and Abediankenari, 2013.
			-	Stomach cancer cell line	Hashemi et al., 2011
		Fruit, leaves, and latex	Ethanol	HeLa	Khodarahmi et al., 2011
		Latex	-	Human glioma (U251), human hepatocellular carcinoma (SMMC7721), and human fetal normal liver (L02) cell lines	Wang et al., 2008
		Latex, fruits, and leaves	Ethanolic	HeLa	Khodarahmi et al., 2011
7.	Antiviral	Latex	Methanol	Herpes simplex (HSV-1)	Lazreg Aref et al., 2011
			-	CpHV-1	Camero et al., 2014
			Ethanol	CpHV-1	Camero et al., 2014
			-	HSV-2	Ay et al., 2018
		Leaves	Aqueous	Hep-2, BHK21 and PRK	Wang et al., 2004

S. no	Activity	Parts used	Extract	Organism used/Cell lines	Reference
8.	Antimicrobial	Leaves	Methanol	Aspergillus brasiliensis, Candida albicans, Bacillus cereus, Bacillussubtilis, Staphylococcus aureus, Enterococcus faecalis, Micrococcus luteus, Klebsiella pneumonia, Pseudomonas aeruginosa, Escherichia coli, and Salmonella sp.	Mahmoudi et al., 2016
			Methanol	S. mutans, S. sanguinis, S. sobrinus, S. ratti, S. criceti, S. anginosus, S. gordonii, Aggregatibacter actinomycetemcomitans, Fusobacterium nucleatum, Prevotella intermedia, Porphyromonas gingivalis, E. coli, S. aureus, S. epidermidis, and S. pyogenes	Jeong et al., 2009
			Ethanol, Methanol, and aqueous	S. aureus, B. cereus, B. subtilis, S. faecalis, K. pneumoniae, P. aeruginosa, S. typhimurium, Aeromonas hydrophila, E. coli, and C. albicans	Keskin et al., 2012
			Ethanol	S. aureus, Streptococcus pyogenes, K. pneumonae, P. aeruginosa, S. typhi, E. coli, C. albicans, Fusarium oxysporum, A. nigar	Rashid et al., 2014
		Fruits	80% methanol, 60% acetone, aqueous, petroleum ether	S. aureus, E. faecalis, B. subtilis subsp. spizizenii, P. aeruginosa, P. aeruginosa, E. coli, E. cloacae, S. enterica subsp. heindelberg, K. pneumonia, and C. albicans	Debib et al., 2014

Table 8. (Continued)

S. no	Activity	Parts used	Extract	Organism used/Cell lines	Reference
		Leaves	Methanol	K. pneumoniae, B. cereus, E. aerogens, B. substilus, S. epidermidus	Ahmad et al., 2013
			Ethanol	S. aureus, E. coli and P. aeruginosa	Weli et al., 2015
			Methanol	S. aureus and E. coli	Rostam et al., 2018
		Latex	Methanol, hexane, chloroform and ethyl acetate	Trichophyton rubrum, T. soudanense, Microsporum canis), A. fumigattus, Scopulariopsis brevicaulis, C. albicans, and Cryptococcus neoformans	Aref et al., 2010
		Bark	Methanol	S. aureus, K. pneumonae, P. aeruginosa, S. typhi, E. coli, B. subtilis, P. fluorescens, S. epidermidis, Staph. haemolyticus, C. albicans, F. oxysporum, Mucor racemosus, Aspergillus nigar, M. mucedo, A. flavus, Rhizopus stolonifer	Azam et al., 2019
				B. subtilis, S.aureus, B. megaterium, P. aeruginosa, E. coli and Proteus vulgaris	AL Yousuf 2012
		Leaves and bark	Dichloromethane and methanol	E. coli, B. subtilis, S. aureus, and P. aeruginosa	Azam Farooq et al., 2013
		Latex, leaves and stem	Methanol	P. aeruginosa, E. coli, S. aureus, and K. pneumoniae	Doro et al., 2018
		Bark	Methanol	Wister albino rats	Bhat et al., 2013
9.	Anti-diabetic	Leaf, peel and pulp	70% ethanol	-	Comparative Antioxidative and Antidiabetic Activities of Ficus Carica Pulp, Peel and Leaf and their Correlation with Phytochemical Contents Iffat Ara

S. no	Activity	Parts used	Extract	Organism used/Cell lines	Reference
		Fruit, leaves and stembark	Ethanol	-	Mopuri and Islam, 2016
		Leaves	Hexane, ethyl acetate and methanol	Albino Wistar rats	Stephen Irudayaraj et al., 2017
			Aqueous	Streptozotocin-induced diabetes rats	Canal et al., 2000
			Methanol	Albino-Wistar rats	Stalin et al., 2012
10.	Hepatoprotective	Leaves	Ethanol	Albino rats	Mujeeb et al., 2011
		Latex	-	Wistar albino rats	Aziz, 2012.
		Leaves	Petroleum ether	Rats	Gond and Khadabadi, 2008
		Stems	Aqueous	Albino Wistar rats	Saoudi and El Feki, 2012
		Leaves	Ethanol	Male albino mice	Aghel et al., 2011
				Albino rats	Mujeeb et al., 2011
11.	Nematicidal	Leaves	Methanol	*Bursaphelenchus xylophilus, Panagrellus redivivus* and *Caenorhabditis elegans*.	Liu et al., 2011
12.	Antihyperlipidemic	Fruits	70:30 (v/v) ethanol/water	Male Wistar rats	Belguith-Hadriche et al., 2016

Conclusion

F. carica is a blessed and essential medicinal plant that has been largely used in many countries as a traditional medicine. In the treatment and prevention of many complications, all parts of this plant were included. The most bioactive compounds in this plant are flavonoids and various extracts have been found to possess biological activity. The potential uses for many illnesses as a medicinal treatment represent less toxicity of this plant. Fig leaves contain polyphenols that are potentially beneficial for human health with antioxidant and radical scavenging effects. Traditionally, fig leaves have been used for treating diabetes and liver disorders.

Chapter 4

Juglans regia L.: Ethnobotany, Pharmacology and Phytochemistry

Abstract

Juglans regia Linn is a regarded as one of the most important medicinal trees that has been widely used in traditional medicine for a wide range of ailments that include helminthiasis, diarrhea, sinusitis, stomachache, arthritis, asthma, eczema, scrofula, skin disorders, and various endocrine diseases such as diabetes mellitus, anorexia, thyroid dysfunctions, cancer and infectious diseases. The present chapter provides comprehensive information on the morphology, systematic position, ethnobotanical uses, pharmacology and phytochemical studies of *J. regia*. Data regarding the current review has been gathered from the books and scientific articles published in various databases such as Science Direct, Web of Science, Scopus, PubMed, and Scientific Information Database.

1. Introduction

Juglans regia, commonly known as Walnut, is a member of family Juglandaceae. The species is a monoecious deciduous tree bearing male and female reproductive organs in separate flowers on the same tree. The walnut is known for its high timber quality and nuts (Thakur, 2011). The tree attains a height of 25-30m and a trunk of up to 2m in diameter. The tree with a short trunk and broad crown is characterized by four-celled nut, a husk which separates from the nut at maturity. The leaves are compound and composed of 7-9 leaflets which have prominent herring-bone venation. Leaflets are ovate with pointed tips and smooth margins (Polunin, 1977).

Catkins (male inflorescences) are borne laterally on one year old tree and pistillate flowers are borne terminally or laterally on the same year in spikes of typically 2 to 3 flowers. The female flowers lack sepals and petals and are pubescent, small and green. The female flowers have an ovary with a large feathery stigma. Male flowers generally appear later in life than female flowers (Westwood, 1977; Jackson and McNeil, 1999).

The pollination in *J. regia* occurs by means of wind and the species is protandrous. Although most of the walnuts are self-fertile, female flowers receive pollens from different plants due to different timing of maturation of male and female flowers. The pollens from the anthers are released 10-12 days after bud break (Jackson and McNeil, 1999). Nuts develop in groups of 1-3 on shoot tips. The nuts are surrounded by green fleshy husk which splits irregularly at maturity. The husk separates from the nutshell easily. The shell is rough, wrinkled and thin. The nut production usually starts when the tree attains age of 4-6 years but commercial production starts after the age of 10 years.

1.1. Taxonomical Description

- **Kingdom:** Plantae
- **Subkingdom:** Viridiplantae
- **Superdivision:** Embryphyta
- **Division**: Tracheophyta
- **Class:** Manoliopsida
- **Order:** Fagales
- **Family:** Juglandaceae
- **Genus:** *Juglans*
- **Species:** *Juglans regia*
- **Synonyms**: *Juglans duclouxiana* Dode, *J. fallax* Dode, *J. kamaonia* C. DC. Dode, *J. orientis* Dode, *J. regia* subsp. *fallax* Popov, *J. regia* var. *kamaonia* C. DC., *J. regia* var. *sinesis* C. DC., *J. sinesis* C. DC. Dode and *Regia maxima* Loudon ex C. DC.

1.2. Vernacular Names

- **Arabic**: Joz
- **Chinese:** Hu tao
- **English**: Walnut, Carpathian walnut, English walnut, Madeira walnut, Persian walnut
- **French:** Noyercommun

- **Indian:** Akhrot
- **Portuguese:** Nogueira-comum
- **Spanish:** Nogalcomúm, Nogaleuropeo
- **Swedish:** Valnöt (Devi et al. 2011)

2. History and Geographical Distribution of *Juglans regia*

J. regia is one of the oldest nuts producing species in the history of mankind. Early history indicates that English walnuts came from ancient Persia where the walnut trees were reserved for royalty, due to this reason it is often known as "Persian walnut." Walnuts were traded along the Silk route between Asia and the Middle East. Caravans used to transport walnuts too far off places through sea trade and ultimately the popularity of the walnuts reached throughout the world. The nut trade business nowadays continues to be well established and structured and the California walnut is famous for its top quality throughout the globe.

The species is being cultivated in most of the temperate parts of the Northern Hemisphere due its high nut quality. The species is reported to grow in the wild across 19 countries in the Asian continent. It occurs from Turkey and Iraq in the West, countries around the Caspian Sea (Azerbaijan, Armenia, Russia and Russia, Iran and Turkmenistan) and further North (Uzbekistan, Kyrgzstan, Tajiskistan and Kazakstan) to the North-eastern Xinjiang region of China. Southwards it is found in Afghanistan and extends south to the Karakoram and Himalayan Mountain ranges of Pakistan, India, Nepal and Bhutan and finally reaches its south eastern limit in Bangladesh, the hills of upper Mynamar and southern China (Gamble, 1902; Zohary, 1993; Kolov, 1998; Hemery, 2000, Thakur, 2006). The species disappeared from the South east of Europe and south west Turkey during the last glaciations and reappeared in the Balkans and the west Turkey in the latter part of second millennium BC.

3. Ethnobotanical Uses

As per the archeological evidence, the consumption of walnuts by humans was started approximately 7300 yr. B.P in proximity to the Mediterranean (Hayes et al. 2015). Historically various parts of the species such as seeds, bark, leaves and green husks were used by people as natural remedies in folk medicine. The seed or kernel which is the edible part of the fruit was consumed in either fresh or in dried form. The species was popular globally and was given great value for its nutritional and health promoting properties (Popovici, 2013). The walnut leaves were extensively used in traditional medicine for the curing of skin infection, for the treatment of sinusitis, stomach ache, ulcers, diarrhea, venous insufficiency, hyperhidrosis haemorrhoidal symtoms, as anti-helmintic, depurative, antioxidants, antiseptic, antibacterial, astringent and chemo preventive purposes (Eidi et al. 2013). The root and stem bark were used as an anti-helmintic, astringent and detergent. The bark of the stem after drying was used as tooth cleaner and whitener. The decoction of leaves and bark was used with alum for staining wool brown (Devi et al. 2011, Kale et al. 2010). In Indian Himalaya, the leaves were being used as mosquito replant.

In Turkish folk medicine, fresh leaves were applied on the naked body or forehead for reducing fever, leaves were also used to overcome the rheumatic pain (Yesilada, 2002). In Iranian traditional medicine, the kernel of the *J. regia* has been used for the treatment of inflammatory bowel disease (Kim et al. 2006). In Palestine, the kernals are used for treating diabetes and asthma as well as for treating prostrate and vascular problems (Jaradat, 2005; Kaileh et al. 2007; Spaccarotella et al. 2008). The tree is used as a remedy for dermal inflammation and excessive perspiration of the hands and feet. The leaves are used to cure scalp itching and dandruff, sunburn and superficial burns (Gruenwald et al. 2001; Robbers and Tyler, 1999; Ali-Shtayehand Abu Ghdeib, 1999; Blumenthal, 2000; Baytop, 1999).

The plant also has a great anti-antherogenic potential and an outstanding osteoblastic activity which leads to beneficial effects of a walnut rich diet on cardio protection and bone loss (Papoutsi et al. 2008). In China, the branches, bark and exocarp of the immature fruit of this tree have been used for treatment of gastric, liver and lung cancer. In the northeastern region of Mexico, the traditional healers used this plant for curing liver problems (Torres-Gonzalez, 2011). The bark paste is used in arthritis, skin diseases, toothache, and hair growth in Nepal. The seed coat is used for healing wounds (Kunwar and Adhikari, 2005). The nut shell of *J. regia* is used in Calabria folk medicine to heal malaria (Tagarelli et al. 2010).

4. Pharmacological Effects

Literature has revealed that the different plant parts exhibited various biological activities which are discussed in detail in the following sections.

4.1. Antibacterial Activity

The aqueous extracts taken from leaves, fruits, barks, and green husks of *J. regia* displayed broad spectrum antibacterial activity against gram-positive and gram-negative bacteria viz., *Bacillus cereus*, *B. subtilis*, *Pseudomonous aeruginosa*, *Escherichia coli*, *Staphylococcus aureus*, *Klebsiella pneumonia*, *Micrococcus luteus*, *Salmonella typhimurium* and *Proteus* species using agar streak method and disc diffusion method (Torres-Gonzalez, 2011; Ibrar, and Sultan, 2007; Kunwar and Adhikari, 2005; Tagarelli et al. 2010; Deshpande et al. 2011).

4.2. Antifungal Activity

The aqueous and other solvent extracts of leaves, fruits and bark of *J. regia* showed antifungal activity against broad range of fungi using disc diffusion method, agar dilution method and Raddish method. According to Pereira et al. (2008), all the walnut varieties exhibited antifungal activity against *Candida albicans* and *Cryotococcus neoformans* when soxleted with petroleum ether. Cold extract of fruit, leaves and bark inhibited the growth of *microsporum canis*, *Trichophyton violaceum* (Ali-Shtayehand Abu Ghdeib, 1999). The aqueous extract of green husks on the other hand showed no antifungal activity against *Candida albicans* and *Cryotococcus neoformans* (Oliveira et al. 2008).

4.3. Anticancer Activity

The walnut milk is considered as a potential anticancer agent. It has been shown that walnut milk drastically reduced cell viability and selectively induces caspase-dependent apoptosis (Doganlar and Doganlar, 2016). The anticancer activity of *J. regia* leaf extracts in hexane and its effects on cell cycle analysis, apoptosis and cancer cell morphology was analyzed in detail in human prostate cancer (PC3) cells.

The extract showed a potent and dose dependent anti-proliferative activity against human prostate cancer cells in vitro. The extract also induces significant apoptosis in PC3 cancer cells as depicted by annexin V binding assays as well as inverted phase contrast microscopy (Li et al. 2015). The cytotoxic effects of *J. regia* extracts and Juglone, a napthoquinone isolated from the chloroform extract of roots and its novel series of triazolyl analogs were studied against various human cancer cell lines. The different extracts of *J. regia* and the Juglone exhibited satisfactory cytotoxic activity against different human cancer cell lines.

The anticancer effect of a methanol extract of walnuts was estimated on human MDA-MB-231, MCF7 and HeLa cells. The extract was found to be cytotoxic to all cancer cells. The extract reduces the intracellular pH, depolarizes the mitochondrial membrane with the release of cytochrome c and flipping of phosphatidylserine. The apoptotic and anti-proliferative activities in chloroform extracts of leaves were examined in human breast and oral cancer cell lines (MCF-7 and BHY).

Bioassay guided fractionation of chloroform extract of *J. regia* leaves lead to the isolation of 5-hydroxy-3, 7, 4 trimethoxyflavone, lupeol, daucosterol, 4-hydroxy-α-tetralone, β-sitosterol and regiolone. All of these compounds inhibited proliferation of MCF-7 (human breast adenocarcinoma) and BHY (Human oral squamous carcinoma) cells in a concentration dependent manner. The effect of green husks of walnut on cell proliferation was evaluated on PC-3 human prostate cancer cells and it was revealed that green husk extracts suppressed proliferation and induced apoptosis in a dose and time dependent manner by modulating expression of apoptosis related genes.

4.4. Antidiabetic Activity

The antidiabetic effect of leaves of *J. regia* in type 1 diabetes was examined in streptozotocin induced diabetes in rats. Treatment with the leaf extracts resulted in a significant decrease in blood glucose, LDL, triglyceride and total amount of cholesterol and a significant increase in insulin and HDL level (Shaw et al. 2010; Mohammadi et al. 2010).

The mechanism of hypoglycemic action of methanolic leaf extracts was analyzed in rats. After 3 weeks of treatment, the plant extract had a significant hypoglycemic action in both short- and long-term models. The antidiabetic effect of ethanolic leaf extract was examined in nondiabetic alloxan-induced diabetic rats. Fasting blood glucose decreased significantly in diabetic rats treated with *J. regia*. The level of insulin increased and glycosylated hemoglobin decreased significantly in diabetic groups receiving *J. regia* extracts compared to those without any treatment. Besides, there was also an increase in the size of islets of Langerhans in *J. regia*-treated rats (Asgary et al. 2008).

The hypoglycemic effect of oral methanolic extracts of leaf and fruit peel of walnut was evaluated in alloxan induced diabetic rats. After four weeks blood was collected for biochemical analysis and pancreas were removed for β-cell counts in histological sections. It was shown that diabetes increased fast blood sugar (FBS) and HbA1c and decreased β-cell number and insulin. FBS reduced only in leaf extract treated group. HbA1c decreased in leaf extract and insulin groups. The β-cell increased in leaf and peel extract groups. Insulin increased moderately in all treatment groups (Javidanpour et al. 2012).

Fifty-eight Iranian male and female patients with type-2 diabetes were enrolled in clinical trials receiving *J. regia* leaf extracts for 2 months for determination of HbA1c and blood glucose level as a main outcome and insulin, SGOT, SGPT and ALP level as secondary outcome. The results revealed that serum fasting HbA1c and blood glucose levels were significantly decreased and the insulin level was increased in patients in the *J. regia* group (Hosseini et al. 2014).

4.5. Antioxidant Activity

The *in vitro* activity of hydro-alcoholic husk extract was analyzed on serum LDL oxidation caused by copper sulphate. The results indicated that the hydro-alcoholic husk extracts reduce serum LDL oxidation (Ahmadvand et al. 2011). The antioxidant activity and protective effects in stabilizing sunflower oil of methanolic extract of *J. regia* green husk were studied. Total phenolics and flavonoids were 144.65±2.1 mg quercetin and 3428.11± 135.80mg gallic acid per 100 gram of dry sample respectively. EC_{50} values of extract in reducing power and DPPH essays were 0.19 and 0.18mg/ml respectively. The 400ppm extract was as effective as 200ppm BHA in retarding sunflower oil deterioration at 600C (Rahimipanah et al. 2010).

The methanolic extract of leaves showed strong antioxidant activity with the use of DPPH radical scavenging and reducing power tests (Hatamjafari and goorabi, 2013). The methanolic extracts of flowers were capable of scavenging H_2O_2 in a concentration dependent manner (Ebrahimzadeh et al. 2013). The different extracts of *J. regia* bark in hot water, hydroalcohol, chloroform and petroleum ether were analyzed for antioxidant activity using different methods including DPPH radical scavenging, ABTS radical scavenging, FRAP assay and superoxide radical scavenging assay. The results showed that the hot water and hydroalcoholic extracts of *J. regia* possessed potent antioxidant activity than the chloroform and petroleum ether extracts.

The ethanolic extract of kernals of *J. regia* showed a protective effect on the DNA strand breaks induced by thiol/Fe^{3+}/O_2 mixed function oxidase, tert-butyl hydroperoxide or UVC radiations in acellular and cellular models (Calcabrini et al. 2017).

4.6. Antiparasitic Effects

The stem and bark extracts of *J. regia* were studied for anthelminthic activity on adult Indian earthworm *Pheretima posthuma*. Different solvents such as petroleum ether, benzene, chloroform, acetone, methanol, ethanol and distilled water for stem and bark extraction. The benzene, ethanol and methanol extracts showed significant anthelminthic activity as compared to that of standard drug piperazine citrate (Vishesh et al. 2010).

The anthelminthic activity of different extracts of leaf was tested against adult Indian earthworm *Pheretima posthuma*. The methanolic leaf extracts demonstrated paralysis as well as death of worms in a less time as compared to piperazine citrate especially at concentration of 50mg/ml. Water extracts showed significant activity while as petroleum ether was least effective among all the extracts (Das et al. 2011).

The antileishmanial activity of *J. regia* hydroalcoholic extract was tested on the growth of the promastigotes of *Leishmania major*. The results showed that *J. regia* extracts reduced the promastigotes number significantly (Serakta et al. 2013).

4.7. Anti-Inflammatory and Analgesic Effects

The ethanolic extracts of leaves of *J. regia* showed potent anti-inflammatory activity against carrageenan induced hind paw edema model in mice without inducing any gastric damage (Erdemoglu et al. 2003). When BV-2 microglial cells were treated with walnut methanolic extract prior to LPS stimulation, the production of nitric oxide and expression of inducible nitric oxide synthase were attenuated. Walnut extract also decreased tumor necrosis-alpha (TNF-alpha) production. The extracts induced internalization of the LPS receptor, tool-like receptor 4 and the anti- inflammatory effects of walnut were dependent on functional activation of phospholipase D2 (Willis et al. 2010).

4.8. Effect on Gastric Ulcer and Colitis System

The gastro protective effect of aqueous extract of *J. regia* leaves was studied in albino rats. The aqueous leaf extract of *J. regia* was investigated for its anti- ulcer activity against pylorus ligation, aspirin induced and ethanol induced gastric ulcer in rats at 500mg/kg bwpo. A significant reduction ($P<0.01$) in ulcer index was seen in leaf extracts of *J. regia* treated rats of pylorus ligation, aspirin induced and ethanol induced gastric ulcer models.

The gastro protective effect was further confirmed by histopathological examination of rat stomach (Dabburu et al. 2012). *J. regia* (walnut) extract was evaluated in acute and chronic murine colitis. In the acute colitis model, mice were given 4% dextran sulfate sodium (DSS) for 5 days. Walnut extract (5 mg/kg/day and 20 mg/kg/day) was dissolved in PBS and administered once daily by oral gavage, beginning 2 days before DSS administration. Severe colitis was induced by DSS administration for 5 days.

In contrast, administration of walnut extract significantly reduced the severity of DSS-induced murine colitis, as assessed by the disease activity index and colon length. In the histopathological analysis, histological grading showed that walnut extract significantly reduced overall colitis score in comparison to the scores of the PBS-treated controls. Immuno-histochemical analysis showed that the DSS-induced phospho-IKK activation in intestinal epithelial cells was significantly decreased in walnut extract–treated mice. Immunoreactivity for occluding was significantly inhibited by the treatments with walnut extract. Walnut extract was also significantly reduced the severity of chronic colitis in mice (Koh et al. 2016).

4.9. α-Amylase Inhibitory Effect

J. regia was tested for α-amylase inhibition. Different concentrations of leaf aqueous extracts were incubated with enzyme substrate solution and the activity of enzyme was measured. Extract showed time and concentration dependent inhibition of α-amylase. 60% inhibition was seen with 0.4 mg/ml of *J. regia* aqueous extract. Dixon plots revealed the type of α-amylase inhibition was competitive inhibition (Rahimzadeh et al. 2014).

4.10. Toxicity and Side Effects

Juglone found in *J. regia* can cause irritation and skin hyper pigmentation in experimental animals but, contact allergy was considered a very rare event in man (Woods and Calnan, 1976; Hausen, 1981). Methanol leaf extract exhibited no toxicity up to 4 g/kg when injected ip in mice (Ravanbakhsh et al. 2016). Mice were given orally 2000 mg/kg walnut polyphenols and maintained for observation for 14 days.

No fatal event occurred, nor abnormal changes observed upon comparison with control group. No evident abnormalities detected in organs upon autopsy. Oral LD50 of walnut polyphenols was deduced to be >2.000mg/kg for both male and female mice (Ali Esmail Al-Snafi, 2018). No liver, kidney and other side effects were observed with the using of 100 mg leaves extract capsules for 3 months in patients, except more GI events (especially a mild diarrhea) associated with extract treatment at the beginning of the study.

5. Phytochemistry of *Juglans regia*

The extensive Phytochemical analysis in different parts of the plants such as roots, leaves, stem, bark, fruit husk, fruits have been carried out by a number of researchers throughout the world which led to the identification of various chemicals such as terpenoids, alkaloids, flavonoids, phenolic acids, fatty acids, cardiac glycosides, steroids, minerals, tannins, protein, dietary fiber, melatonin, plant sterols, α- tocopherol, folate, tannins, vitamin A and C, and vitamin E family compound and. The isolated Compounds are listed in Table 9.

Table 9. Classes and names of compounds isolated from the various parts of *J. regia*

S. No.	Part of plant	Classes	Compounds	References
1.	Male flower	Phenols and Flavonoids	Artemisinin, ursolic acid, 4,7,10,13,16,19- Docosahexaynoic acid (omega-3 fatty acid), Elephantopin, gamma-L-Glutamyl-L cysteine, Pantothenic Acid (vitamin B5), 3,4,5-trihydroxy benzoic acid (Gallic acid), Indole-3-ethanol, Quercetin-3-O-rutinoside, Securinine, Decanoic acid, Madecassic Acid, Dihydromyricetin, (9Z,12Z)-9,12-Octadecadienoic acid (Linoleic acid), Dihydromyricetin (Ampelopsin), C16 Sphinganine, Quercetin-3-O-glucuronide (Miquelianin), (9Z)-Octadecenoic acid (Oleic acid), Swietenine, n-Hexadecanoic acid (Palmitic acid)	Muzaffer and Paul (2018)
2.	Green husk, leaves, stem bark and nuts	Phenolic Compounds	2,2-Dimethyl-3-OxoButyric Acid 2-Trimethylsilanyl, Quercetin-3-O-glucuronide, Hexanoic acid, trimethylsilyl ester	Nabavi et al. (2011); Kale et al. (2012); Anjum et al. (2016).
3.	Walnut seeds	Phenolic Compounds	Oleic acid, n-Hexadecanoic acid, 1,2–Benzene dicarboxylic acid, Ellagic acid, p-coumaric acid, Syringic acid, 5-Ocaffeoylquinic., Ferulic acid	Colaric et al. (2005); Fukuda et al., (2003)
4.	Walnut leaves	Phenolic compounds	3-caffeoylquinic acid and, 3-p-coumaroylquinic acid, 4-pcoumaroylquinic acid, Quercetin 3-galactoside, Quercetin 3-arabinoside, Quercetin 3-xyloside, Quercetin 3-rhamnoside, Quercetin 3-pentoside, Kaempferol 3-pentoside, 3-O-caffeoylquinic acid, 3-O-p-coumaroylquinic acid, 4-O-p-coumaroylquinic acid	Ahmed (2015); Amaral et al. (2004); Pereira et al. (2007)
	Walnut leaves	Flavonoids	5-hydroxy-3,7,4'trimethoxyflavone, lupeol, daucosterol, 4-hydroxy-α-tetralone, β-sitosterol, 5,7- dihydroxy-3,4'dimethoxyflavone regiolone, syringetin-o-hexoside, myricetin-3-o-glucoside, myricetin-3-o-pantocid, aesculetin, taxifolin-pantocid, quercetin glucuronide, kaempferol pantocid, kaempferol and rhamnoside.	Salimi et al. (2014)

Table 9. (Continued)

S. No.	Part of plant	Classes	Compounds	References
	Husks	Cyclic diarylheptanoids	Rhoiptelol, Juglanin A, Juglanin B, Juglanin C	Li et al. (2008); Liu et al. (2008)
	Seeds	Tannins	1, 2, 3, 4, 6-penta-O-galloyl-3-D-glucose, Rugosin C, 1, 2, 3, 6-tetra-O-galloyl-3-D-glugose Tellimagrandin II, Casuarictin 1-degalloylrugosin F	Jin and Qu (2008)
	Green walnut liqueur	Phenolic compounds	Protocatechuic, Sinapic and p-coumaric acids as well as 1,4-naphthoquinone, chlorogenic acid, caffeic acid, ferulic acid, sinapic acid, gallic acid, ellagic acid, protocatechuic acid, syringic acid, Salicylic acid, Rutin, Myricetin, vanillic acid, catechin, epicatechin, myricetin, 1,4-naphthoquinone, juglone	Jakopic et al. (2008); Stampar et al. (2006); Cosmulescu et al. (2014)
	Walnut seed coat	Phenolic compounds	Juglone, syringic, and ellagic acid	Colaric et al. (2005)

Conclusion

The current chapter highlights the systematic position, distribution, history, traditional uses, pharmacology and phytochemistry of *J. regia*. Although lot of research has been carried out in the past on the plant species but still there are some areas which needs to be focused on for further elucidating the biochemical pathway and other pharmacological activities which are important for future drug discovery and development.

Chapter 5

Morus alba L.: Traditional Uses, Phytochemistry, and Pharmacology

Abstract

This chapter presents comprehensive phytochemical, pharmacological and ethnobotanical information about *Morus alba*. The species serves as a multifunctional factory producing food, cosmetics, medicine, and fodder. The current review on the pharmacological reports indicates that the herbal uses of the plant are scientifically valid in diabetes, hypertension, central nervous system (CNS) problems, hepatoprotective and cardiac diseases. Some important traditional uses about this species are vanishing with age due to lack of documented proof, and the information regarding its medicinal use is only limited to those who are familiar with the species. Further research on the plant species will boost the discovery of new bioactive compounds which will disclose many important pharmacological activities of the species.

1. Introduction

Mulberry belongs to the *Morus* genus of the Moraceae family and it is located in East, West, and Southeast Asia, South Europe, South of North America, Northwest of South America, and several regions of Africa (Pel et al., 2017; Watson and Dallwitz, 2007). There are twenty-four species of *Morus*, and at least 100 varieties of recognized subspecies that adapt to different climatic, topographical, and soil conditions (Ercisli and Orhan, 2007).

M. alba is called "white mulberry/silkworm mulberry" due to the color of the fruit and bark *M. alba* species were generally thought to have evolved in China and India at the bottom of the Himalayas (Awasthi et al., 2004), so assessing their genetic framework is considerably more essential in the Himalayas and its adjacent regions, as unveiled by RAPD and ISSR marker assays (Awasthi et al., 2004). In India, mulberry is mainly present in Andhra Pradesh, Assam, Bihar, Delhi, Haryana, Himachal Pradesh, Jammu and

Kashmir, Karnataka, Kerala, Madhya Pradesh, Maharashtra, Manipur, Meghalaya, Punjab, Rajasthan, Tamil Nadu, Uttar Pradesh, Uttarakhand and West Bengal (Tikader, 2011).

The species is fast growing in nature, and it may reach a height of up to 20 meters. For cultivation purposes, the trees are reduced to a low-growing bush to aid the harvesting of leaves or fruits. Bark is grey-brown in color with horizontal lenticels. The leaves are glittering green, alternate, cordate at the base and acuminate at the top, the margins are serrated, and the petioles are long and thin. The leaves are variable in shape, varying from 5.0−7.5 cm in length. Even on a single tree, some are unlobed while others are mostly palmate.

In the temperate and sub-tropical areas, the trees are frequently dioecious but may be monoecious, and occasionally can change from one sex to another. The flowers comprising male and female catkins are inconspicuous, pendulous, and greenish. The fruit consists of numerous drupes formed by individual flowers to form a sorosis, the characteristic mulberry fruit. The color of the fruit is green when small, turns orange-red, and eventually to purplish- black when fully ripe.

1.1. Taxonomical Description

- **Kingdom:** Plantae
- **Subkingdom:** Tracheobionta-Vascular plant
- **Superdivision:** Spermatophyta
- **Division:** Magnoliphyta-Flowering plant
- **Class:** Magnoliopsida-Dicotyledon
- **Subclass:** Hamamelididae
- **Order:** Urticales
- **Family:** Moraceae-Mulberry family
- **Genus:** *Morus*
- **Species:** *Morus alba*
- **Synonyms:** *Morus alba* f. alba, *Morus alba* var. *alba*, *Morus alba* var. *atropurpurea* (Roxb.) Bureau, *Morus alba* var. *bungeana* Bureau, *Morus alba* var. *laevigata* Wall. ex *Bureau*, *Morus alba* var. *latifolia* (Poir.) Bureau, *Morus alba* var. *mongolica* Bureau, *Morus alba* var. *multicaulis* (Perr.) Loudon, *Morus alba* var. *nigriformis* Bureau, *Morus alba* var. *serrata* (Roxb.) Bureau, *Morus alba* var.

stylosa Bureau, *Morus alba* var. *tatarica* (L.) Loudon, *Morus alba* var. *tatarica* (L.) Ser.

1.2. Vernacular Names

- **English**: Mulberry, White mulberry
- **Arabic**: Al tooth
- **Chinese**: Bai sang, hong sang, sang bai, sang Zhi
- **Greek**: Aspri moria
- **Italian**: Gelso bianco
- **Japanese**: Kuwa, Guwa, Ma guwa
- **Pakistani**: Tut, tut kishmishmi, tutri
- **Unani**: Shahtuut, Tuut
- **Indian**: Toot, chimmi, shehtoot, tutri
- **Sansakrit**: Toota, Brahmataru, Tooda
- **Tamil**: Kambli Chedi
- **Mizo**: Thing-theihmu
- **Manipuri**: Kabrangchak angouba
- **Oriya**: Tuto, Tuticoli
- **Punjabi**: Tut, Tutri
- **Kannada**: Bili uppu nerale
- **Telegu**: Reshme chettu, Pippalipandu chettu
- **Gujarati**: Shetur
- **Marathi**: Tut, Ambat
- **Bengali**: Tut

2. Traditional Uses

Mulberry plants are considered a staple food of silkworms, which has supported the silk industry for centuries (Sanchez, 2002). The silkworm takes only mulberry leaves to produce its cocoon, which generates the silk, and there is a strong relationship between the content of leaf protein and the performance of cocoon creation (Machii et al., 2000). The amino acids (threonine, valine, methionine, leucine, phenylalanine, lysine, histidine, and arginine) present in mulberry leaves are essential for silkworm growth. It has been recognized that the growth of silkworms and the quality of cocoon and raw silk quality rely on the grade of mulberry leaves, which are closely linked to the plant varieties, environmental conditions, and cultivation practice (Kumar et al., 2014). In China, 15–18 kg of mulberry leaves is required to make 1 kg of the cocoon at the farm level (Chan et al., 2016).

Mulberry leaves are also used as fodder livestock feed. They are nutritious, palatable, and non-toxic and can increase milk production when given to dairy animals (Arabshahi-Delouee and Urooj, 2007). Mulberry leaves have higher protein content and better biodegradable digestion than most tropical grasses used for fodder (Vu et al., 2011).

In traditional Chinese Medicine (TCM), *M. alba* leaves, fruits and bark have long been used to treat fever, protect liver damage, enhance eyesight, reinforce joints, help with urination, and reduce blood pressure (Bae and Suh, 2007). The Chinese Ministry of Health selected mulberry as one of the first and foremost medicinal and edible plants in 1985, and its therapeutic use was noted in the Chinese Pharmacopoeia (Chinese Pharmacopoeia Commission, 2015).

Historically, *M. alba* fruit juice has been commonly used in jaundice and liver problems in various parts of Pakistan (Abbasi et al., 2009). Patients with diabetes in Korea and Japan eat mulberry leaves as an anti-hyperglycemic supplement (Katsube et al., 2006). Mulberry leaves are effective in lowering blood sugar levels associated with high blood pressure, alcohol, and the hangover from diabetes (Yang et al., 2012). In East and Southeast Asia, the consumption of mulberry tea is gaining popularity. The tea is rich in γ-aminobutyric acid (2.7 mg g-1 dry weight) that is 10-fold greater than that of green tea (Morrison, 2006). The compound is known to lower blood pressure.

In Turkey and Greece, *M. alba* trees are grown for fruits as opposed to foliage (Ercisli and Orhan, 2007). The fruits are used to make mulberry juice, jam, liquor, and canned mulberries. In China, the leaves of *M. alba* are processed into tea while fruit juice is taken as a health beverage (Pothinuch and Tongchitpakdee, 2019). Other uses of mulberry include paper and mushroom production (Rohela et al., 2020). The woodchips of mulberry trees are used as pulp for paper production and as the medium for mushroom culture. In India, mulberry wood is used to make sports equipment, furniture, household utensils, and farm equipment (Sanchez, 2002).

3. Nutritional Value

The mulberry fruits are of great significance in human nutrition. It harbors considerable proteins, lipids, carbohydrates, fiber, minerals and vitamins yet low in calories, which seem to be a good option of food for buyers. Fresh mulberry fruits are capable of providing 1.44 g of protein per 100 g, greater than strawberries (0.67 g/100g) and raspberries (1.20 g/100 g) and similar to blackberries (1.39 g/100 g) (Giampieri et al., 2012; Koum et al., 2012; Rao and Snyder, 2010). A maximum of eighteen amino acids are found in mulberry fruits, as well as nine essential amino acids that human beings require. The ratio of essential amino acids (EAA) to total amino acids (TAA) is 42%, which is similar to some high-quality protein foods for instance milk and fish (Jiang and Nie, 2015). According to FAO/WHO/UNU, the essential amino acid score (EAAS) of many EAAs is greater than 100, which fits the necessities of children or adults (Joint, 1985). Thus, mulberry fruit is a healthy source of protein to some degree.

The sugar content of mulberry fruit increases when ripe (Lin and Lay, 2013). The major sugars are glucose and fructose (Gundogdu et al., 2011; Mahmood et al., 2012). Mulberry fruits are also mineral-rich. The content of macro-elements on dry weight basis is: Nitrogen (1.62-2.13 g/100 g), Phosphorous (0.24–0.31 g/100 g), Potassium (1.62-2.13 g/100 g), Calcium (0.19–0.37 g/100 g), Sodium (0.01 g/100 g), Magnesium (0.12–0.19 g/100 g), Sulphur (0.08–0.11 g/100 g); and micro-elements is: Iron (28.2-46.74 mg/kg), Copper (4.22-6.38 mg/kg), Boran (13.78-19.48 mg/100 g), Manganese (12.33-19.38 mg/kg), Zinc (14.89-19.58 mg/kg) and Nickel (1.40-2,62 mg/kg) (Sánchez-Salcedo et al., 2015b). Furthermore, other vitamins, namely thiamin, riboflavin, niacin, folate, vitamin A, vitamin B-6, vitamin E and vitamin K, are also present in mulberry fruit. Yang et al. (2010) found four tocopherols in the mulberry fruit; γ-Tocopherol (0.245 mg/g, counting β-tocopherol) was found in the greatest quantity, accompanied by δ-tocopherol (0.074 mg/g) and α-tocopherol (0.004 mg/g).

The mulberry seeds produce 25-35% of yellow oil enriched with precious fatty acid group omega-3 (Łochyńska and Oleszak, 2011). Of the 13 recorded acids in the fatty acid test, seven are mono- and poly-unsaturated fatty acids (MUFAs and PUFAs), which belong to the most valuable group of fatty acids, omega-3. Factors, particularly environmental conditions and genetic factors, affect the fatty acid profile (Liang et al., 2012; Sánchez-Salcedo et al., 2016). The major fatty acids in mulberry fruit are usually linoleic acid (C18:2), palmitic acid (C16:0), and oleic acid (C18:1). Linoleic acid (C18:2), that is essential for human growth, health promotion, and disease prevention, was the key fatty acid and should be acquired from outside sources since it cannot be synthesized by humans (Whelan, 2008).

4. White Mulberry in the Food Industry

The benefits of white mulberry leaves and fruits are increasingly being used for food production that affects human health. Besides, the nutritional benefits of mulberry, health status and well-being and/or disease reduction, they are also used in functional food production. Nowadays, an innovative health care diet with mulberry extract has been developed to enhance human immunity and health (Butt et al., 2008; Sánchez-Salcedo et al., 2015b). Polyphenol-rich chocolate, probiotic, and prebiotic chocolate or chocolate with mulberry extract or anthocyanins were studied and produced (Gültekin-Özgüven et al., 2016).

Nowadays, dried and powdered mulberry leaves or their extracts are utilized in various ways. The tea infusion contains no caffeine and has a mild and pleasant taste. The infusion of the leaves reveals very good benefits when applied during flu, cold, and throat inflammations. The dried material is utilized in Indian cooking to make bread from wheat flour called "paratha" and in for rice snacks with dried leaves of white mulberry in Thailand. Furthermore, the white mulberry fruits are added to muesli, which not only boosts the product's antioxidant potential, but also reduces the product's advanced oxidative changes (Kobus-Cisowska et al., 2013).

New functional yogurts using jam and leaves of mulberry have also been reviewed (Lee and Hong, 2010). The use of mulberry fruit is also used to make jams, ice-creams, vinegar, juices, wine, and cosmetic products (Natić et al., 2015). Compared to grape wine, mulberry wine has the higher antioxidant capacity and a better phenolic profile (Celep et al., 2015). The white mulberry fruits contain 20% sugars, largely glucose, maltose, sucrose, and fructose (Ercisli and Orhan, 2007). In addition, they contain organic acids: citric and cider acids and volatile oils. Due to the high content of easily digestible sugars, the populations of the high mountain area mulberry fruits are dried and ground into flour and added to various dishes.

M. alba has very valuable activities for colon microflora *Bifidobacterium* and *Lactobacillus* (Cummings and Macfarlane, 2002). One way to increase the number of intestinal microorganisms is the use of prebiotic agents, which are required to create a bifidogenic effect on the number of prebiotic cultures in the large intestine (Gibson and Fuller, 2000). The oligosaccharides related with intestinal bacteria lead to the maturation of T lymphocytes, which are directly related to the prevention of gastrointestinal diseases (Rolim, 2015). In addition, dried mulberry fruit is used as a food source for regenerative and fortification food (Jeszka et al., 2009). Flavonoids, alkaloids, and stilbenoids are part of the root bark of mulberry. It possesses antimicrobial, skin-whitening, cytotoxic, anti-inflammatory, and anti-hyperlipidemic properties (Chan et al., 2016).

5. Phytoconstituents of Various Parts of *Morus alba*

Extensive phytochemical chemical research has been carried out which led to the isolation of various secondary metabolites such as terpenoids, alkaloids, flavonoids (including chalcones and anthocyanins), phenolic acids, stilbenoids, and coumarins in *M. alba*. Table 10 lists the compounds derived from the leaves, fruit, root, bark, root bark, twig and stem of the plant.

6. Pharmacological Uses

Many pharmacological studies have been reported from the different extracts of *M. alba*. The different parts of the plant show several pharmacological activities including antidiabetic, antiplaque, antiviral, hypolipidemic, anti-inflammatory, antimicrobial, cytotoxic, hypotensive, vascular protective, anti-obesity, hepatoprotective, antiplatelet, anthelmintic, anxiolytic and antioxidant activities. A summary of the reported biological activities is given in Table 11.

Conclusion

The chapter provides in-depth information on traditional uses, phytochemistry and pharmacology of *M. alba*. In addition, the chapter also highlights the botanical aspects, distribution, nutritional values, and role of *M. alba* in food industry. The species is indispensable in silk industry and needs a large-scale cultivation for more production of silk and other raw products. Being a rich source of phytochemicals, further research is needed to unravel the other biological compounds which will be beneficial for drug discovery and development.

Table 10. Classes and names of compounds isolated from the various parts of *M. alba*

S. No.	Part of plant	Classes	Compounds	References
1.	Root bark	Terpenoids	Betulinic acid	(Yang et al., 2011)
			Ursolic acid	
			Uvaol	
		Flavonoids	Oxydihydromorusin	(Nomura et al., 1978)
		Stilbenoids	Alabafuran A	(Yang et al., 2011)
			Artoindonesianin O	
			3',5'-Dihydroxy-6-methoxy-7-prenyl-2-arylbenzofuran	
			Mulberrofurans L, Y	
			Oxyresveratrol 2-O-β-D-glucopyranoside	
		Coumarins	5,7-Dihydroxycoumarin 7-O-β-d-apiofuranosyl-(1→6)-O-β-D-glucopyranoside	
			5,7-Dihydroxycoumarin 7-O-β-D-glucopyranoside	
2.	Roots	Alkaloids	Calystegins B2, C1	(Asano et al., 1994)
			1-Deoxynojirimycin	
			1,4-Dideoxy-1,4-imino-pD-arabinitol	
			1,4-Dideoxy-1,4-imino-pD-ribitol	
			1,4-Dideoxy-1,4-imino-(2-O-β-pD-glucopyranosyl)-d-glucopyranosyl)-D-arabinitol	
			Fagomine	
			3-epi-Fagomine	
			2-O-α-D-Galactopyranosyl-1-deoxynojirimycin	
			6-O-α-D-Galactopyranosyl-1-deoxynojirimycin	
			2-O-α-D-Glucopyranosyl-1-deoxynojirimycin	
			3-O-α-D-Glucopyranosyl-1-deoxynojirimycin	
			4-O-α-D-Glucopyranosyl-1-deoxynojirimycin	
			2-O-β-D-Glucopyranosyl-1-deoxynojirimycin	
			3-O-β-D-Glucopyranosyl-1-deoxynojirimycin	
			4-O-β-D-Glucopyranosyl-1-deoxynojirimycin	

S. No.	Part of plant	Classes	Compounds	References
3.	Fruits		6-O-β-D-Glucopyranosyl-1-deoxynojirimycin	
			N-Methyl-1-deoxynojirimycin	(Hisayoshi et al., 2004)
		Flavonoids	7, 2', 4', 6'-Tetrahydroxy-6-geranylflavanone	(Kusano et al., 2002)
		Alkaloids	2α, 3β-Dihydroxynortropane	
			2β, 3β-Dihydroxynortropane	
			3β, 6exo-Dihydroxynortropane	
			2-[2-Formyl-5-(hydroxymethyl)-1-pyrrolyl-[3-methyl pentanoic acid lactone	(S. B. Kim et al., 2014)
			4-[Formyl-5-(hydroxymethyl)-1H-pyrrol-1-yl] butanoate	
			4-[Formyl-5-(methoxymethyl)-1H-pyrrol-1-yl] butanoic acid	
			2-(5'-Hydroxymethyl-2-formylpyrrol-1'-yl)-3-(4-hydroxyphenyl) propionic lactone	
			2-(5'-Hydroxymethyl-2-formylpyrrol-1'-yl)-3-phenylpropionic acid lactone	
			2-(5-Hydroxymethyl-2', 5'-dioxo-2', 3', 4', 5'-tetrahydrobipyrrole) carbaldehyde	
			2-(5-Hydroxymethyl-2-formylpyrrol-1-yl) isovaleric acid lactone	
			2-(5-Hydroxymethyl-2-formylpyrrole-1-yl) propionic acid lactone	
			2-(Hydroxymethyl-2-formylpyrrole-1-yl) isocaproic acid lactone	
			Methyl 2-[2-formyl-5-(methoxymethyl)-1H-pyrrol-1-yl]-3-(4-hydroxyphenyl) propanoate	
			Methyl 2-[2-formyl-5-(methoxymethyl)-1H-pyrrole-1-yl] propanoate	
			Morroles B-F	
			2α, 3β, 4α-Trihydroxynortropane	(Kusano et al., 2002)
			2α, 3β, 6exo-Trihydroxynortropane	
		Flavonoids	Epigallocatechin	(Natić et al., 2015)
			Epigallocatechin gallate	
			Gallocatechin	
			Gallocatechin gallate	
			Isorhamnetin glucuronide	
			Isorhamnetin hexoside	
			Isorhamnetin hexosylhexoside	

Table 10. (Continued)

S. No.	Part of plant	Classes	Compounds	References
			Kaempferol glucuronide	
			Kaempferol hexoside	
			Kaempferol hexosylhexoside	
			Kaempferol rhamnosylhexoside	
			Morin	
			Naringin	
			Quercetin glucuronide	
			Quercetin hexoside	
			Quercetin hexosylhexoside	
			Quercetin	
		Anthocyanins	Cyanidin 3-O-glucoside	(Qin et al., 2010)
			Cyanidin 3-O-rutinoside	
			Cyanidin 3-O-β-D-galactopyranoside	(Du et al., 2008)
			Cyanidin 3-O-β-D-glucopyranoside	
			Cyanidin 7-O-β-D-glucopyranoside	
			Cyanidin galloylhexoside	(Natić et al., 2015)
			Cyanidin hexoside	
			Cyanidin hexosylhexoside	
			Cyanidin pentoside	
			Cyanidin 3-O-(6′′-O-α-rhamnopyranosyl-β-D-galactopyranoside)	(Du et al., 2008)
			Cyanidin 3-O-(6′′-O-α-rhamnopyranosyl-β-D-glucopyranoside)	
			Cyanidin rhamnosylhexoside	(Natić et al., 2015)
			Delphinidin acetylhexoside	
			Delphinidin hexoside	
			Delphinidin rhamnosylhexoside	
			Pelargonidin 3-O-glucoside	(Qin et al., 2010)
			Pelargonidin 3-O-rutinoside	

S. No.	Part of plant	Classes	Compounds	References
			Pelargonidin hexoside	(Natić et al., 2015)
			Pelargonidin rhamnosylhexoside	
			Petunidin rhamnosylhexoside	
		Phenolic acids	Caffeic acid	(Sánchez-Salcedo et al., 2015b)
			Chlorogenic acid (5-CQA)	(Hisayoshi et al., 2004)
			3-O-Caffeoylquinic acid	
			Ellagic acid	
			Gentisic acid	
			p-Hydroxybenzoic acid	(Dat et al., 2010; Hisayoshi et al., 2004)
			Hydroxyphenylacetic acid methyl ester	(Dat et al., 2010)
			Jaboticabin	
			Kaempferol 3-O-rutinoside	(Sánchez-Salcedo et al., 2015b)
			Methyl 3-O-caffeoylquinate	(Hisayoshi et al., 2004)
			Methyl 4-O-caffeoylquinate	
			Methyl 5-O-caffeoylquinate	
			Methyl dicaffeoylquinate	
			Neo-chlorogenic acid (3-CQA)	(Sánchez-Salcedo et al., 2015b)
			Protocatechuic acid ethyl ester	(Dat et al., 2010)
			Protocatechuic acid methyl ester	
			Quercetin 3-O-glucoside	(Sánchez-Salcedo et al., 2015b)
			Quercetin 3-O-rutinoside	
4.	Leaves	Chalcones	Morachalcones B, C	(Y. Yang et al., 2010b)
		Flavonoids	Astragalin	(Doi et al., 2001)
			Atalantoflavone	(Dat et al., 2010)
			Benzyl d-glucopyranoside	(Doi et al., 2001)
			Cyclomulberrin	(Dat et al., 2010)
			Dihydrokaempferol 7-O-β-D-glucopyranoside	(Wang et al., 2013)
			3′, 8-Diprenyl-4′, 5, 7-trihydroxyflavone	(Dat et al., 2010)

Table 10. (Continued)

S. No.	Part of plant	Classes	Compounds	References
			Euchrenone a7	(Z. Yang et al., 2012)
			8-Geranylapigenin	(Dat et al., 2010)
			7-Hydroxyl-8-hydroxyethyl-4′-methoxylflavane-2′-O-β-D-glucopyranoside	(Z. Yang et al., 2012)
			8-Hydroxyethyl-7, 4′-dimethoxylflavane2′-O-β-D-glucopyranoside	
			Isoquercitrin	(Doi et al., 2001; Katsube et al., 2006; Kim et al., 2000)
			Kaempferol	(Dat et al., 2010; Z. Yang et al., 2012)
			Kaempferol 3-O-β-D-rutinoside	(Wang et al., 2013)
			Kaempferol 3-O-β-D-glucopyranoside	
			kaempferol-3-O-(6″-O-acetyl)-beta-D-glucopyranoside	(S. Y. Kim et al., 1999)
			kaempferol-3-O-alpha-L-rhamnopyranosyl-(1→6)-beta-D-glucopyranoside	(S. Y. Kim et al., 1999)
			7-Methoxyl-8-ethyl-2′, 4′-dihydroxylflavane-2″-O-β-D-glucopyranoside	(Z. Yang et al., 2012)
			7-Methoxyl-8-hydroxyethyl-2′, 4′-dihydroxylflavane	
			7-Methoxyl-8-hydroxyethyl-4′-hydroxylflavane-2′-O-β-D-glucopyranoside	
			Norartocarpetin	
			Quercetin 3, 7-di-O-β-D-glucopyranoside	(Wang et al., 2013)
			Quercetin 3-O-(6″-O-acetyl)-β-D-glucopyranoside	
			Quercetin 7-O-β-D-glucopyranoside	
			Quercetin-3-O-β-D-glucopyranosyl-(1→6)-β-D-glucopyranoside	(S. Y. Kim et al., 1999)
			Quercetin-3,7-di-O-β-D-glucopyranoside	(S. Y. Kim et al., 1999)
			Quercetin-3, 7-di-O-β-D-glucopyranoside	(Kim et al., 2000)
			quercetin-3-O-α-L-rhamnopyranosyl-(1→6)-β-D-glucopyranoside	(S. Y. Kim et al., 1999)
			Quercetin-3-O-beta-D-glucopyranoside	(S. Y. Kim et al., 1999)
			Quercetin 3-(6-malonylglucoside)	(Katsube et al., 2006)
			Roseoside	(Doi et al., 2001)
			Scopolin	

S. No.	Part of plant	Classes	Compounds	References
			Skimmin	
			5, 7, 3'-Trihydroxy-flavanone-4'-O-β-D-glucopyranoside	(Wang et al., 2013)
			5, 7, 4'-Trihydroxy-flavanone-3'-O-β-D-glucopyranoside	
		Stilbenoids	Chalcomoracin	(Z. Yang et al., 2012)
		Polyphenols	1-Caffeoylquinnic acid	(Thabti et al., 2012)
			Caffeic acid	
			5-Caffeoylquinnic acid	
			4-Caffeoylquinnic acid	
			Kaempferol-3-O-glucopyranosyl-(1,6)-β-D-glucopyranoside	
			Kaempferol-3-O-(6-malonyl)glucoside	
			Kaempferol-7-O-glucoside	
			Quercetin-3-O-rhamnoside -7-O-glucoside	
			Quercetin-3,7-D-O-β-D-glucopyranoside	
			Quercetin-3-O-(6-malonyl)-β-D-glucopyranoside	
			Quercetin-3-O-glucoside	
			Quercetin-3-O-glucoside-7-O-rhamnoside	
5.	Leaves and Fruits	Flavonoids	Rutin	(Katsube et al., 2006; Natić et al., 2015; Wang et al., 2013)
		Phenolic acids	5-O-Caffeoylquinic acid	(Hisayoshi et al., 2004; Memon et al., 2010)
			m-Coumaric acid	(Memon et al., 2010)
			p-Coumaric acid	(Hisayoshi et al., 2004; Memon et al., 2010)
			Ferulic acid	
			Gallic acid	
			Hydroxybenzoic acid	
			Protocatechuic acid	(Dat et al., 2010; Hisayoshi et al., 2004; Memon et al., 2010)
			Protocatechuic aldehyde	(Memon et al., 2010)
			Syringaldehyde	
			Syringic acid	

Table 10. (Continued)

S. No.	Part of plant	Classes	Compounds	References
			Vanillic acid	(Dat et al., 2010; Hisayoshi et al., 2004; Memon et al., 2010)
6.	Twigs	Flavonoids	6-Geranylapigenin	(Zhang et al., 2009)
			6-Geranylnorartocarpetin	
7.	Stem	Stilbenoids	Dihydromorin	(Rivière et al., 2014)
			2,3-trans-Dihydromorin-7-O-β-glucoside	
			Steppogenin	
		Coumarins	Isoscopoletin 6-(6-O-β-apiofuranosyl-β-glucopyranoside)	
8.	Bark	Coumarins	5,7-Dihydroxycoumarin 7-(6-O-β-D-apiofuranosyl-β-d-glucopyranoside)	(Piao et al., 2009)
			6,7-Dihydroxycoumarin 7-(6-O-α-rhamnopyranosyl-β-D-glucopyranoside)	
9.	Leaves and Root bark	Flavonoids	Cyclomorusin	(Dat et al., 2010; Doi et al., 2001; NOMURA et al., 1976)
			Kuwanons A–C, E, G, H, J, S, T	(Dat et al., 2010; Nomura et al., 1978; Yang et al., 2011)
			Morusin	(Dat et al., 2010)27
			Sanggenons F, J, K	(Dat et al., 2010; Yang et al., 2011)
10.	Leaves, Stem, and Root bark	Stilbenoids	Moracins C, D, M–P, R, V–Y	(Rivière et al., 2014; Y. Yang et al., 2010a; Yang et al., 2011; Z. Yang et al., 2012)
			Mulberrosides A, B, F	(Lee et al., 2002; Rivière et al., 2014; Yang et al., 2011)
11.	Leaves, Fruits, and Twigs	Flavonoids	Quercetin	(Natić et al., 2015; Wang et al., 2013; Z. Yang et al., 2012; Zhang et al., 2009)
12.	Twigs and Stem	Stilbenoids	Resveratrol	(Jin, 2002; Rivière et al., 2014)
			Oxyresveratrol	

Table 11. An overview of the modern pharmacological assessments of *M. alba*

S. no	Activity	Parts used	Extract	Organism used/Cell lines	Reference
1.	Antidiabetic activity	Root bark	Ethanol	Male Wistar rats	(Singab et al., 2005)
		Root bark	Methanolic	Streptozotocin-induced hyperglycemic mice	(E.-S. Kim et al., 1999)
		Leaves	Acetone and ethanol	Male Wistar rats	(Jeszka-Skowron et al., 2014)
		Fruits	Ethanol	Male Kunming mice	(Wang et al., 2013)
		Fruits	Aqueous	Male Wistar rats	(Jiao et al., 2017)
2.	Antiplaque	Leaves	Hot water	*Streptococcus mutans*	(Lokegaonkar and Nabar, 2011)
3.	Antiviral	Root bark	Ethanol	HSV-1 (15577) strain and vero cells	(Du et al., 2003)
4.	Hypolipidemic	Root bark	Ethanol	Male Wistar rats	(Hesham et al., 2005)
5.	Anti-inflammatory	Stem	Ethanol	RAW 264.7	(Soonthornsit et al., 2017)
		Root bark	Methanol	RAW264.7	(Eo et al., 2014)
		Stem	Ethanol	RAW 264.7	(Wongvat et al., 2020)
6.	Antimicrobial	Root bark	Acetonitrile and water	*Candida albicans, Saccaromyces cerevisiae, Escherichia coli, S. typhimurium, Staphylococcus epidermis,* and *S. aureus*	(Sohn et al., 2004)
		Leaves	Petroleum ether, chloroform, and methanol	*Pseudomonas aeruginosa* (NCIM 2945), *Proteus vulgaris* (NCIM 2027), *Bacillus subtilis* (NCIM 2920), *Salmonella typhi* (NCIM 2501), *Shigella flexneri* (NCIM 4924), *C. albicans* (NCIM 3103), and *Aspergillus niger* (NCIM 789)	(Aditya et al., 2012)
		Leaves, stem, and fruits	Aqueous	*B. subtilis, S. aureus, S. epidermidis, E. coli, P. aeruginosa,* and *C. albican*	(Wang et al., 2012)

Table 11. (Continued)

S. no	Activity	Parts used	Extract	Organism used/Cell lines	Reference
		Leaves	Ethanol and distilled water	*E. coli* (ATCC 25922), *S. aureus* (ATCC 25923), *Enterococcus faecalis* (ATCC 29212), *P. aeroginosa* (ATCC 27853), *C. albicans* (ATCC 10231)	(Emniyet et al., 2014)
		Stem	Ethanolic	*Porphyromona gingivalis* (ATCC 33277) and *Aggregatibacter actinomycetemcomitans* (ATCC 29523)	(Yiemwattana et al., 2018)
		Fruits	Ethanolic	*C. albicans, A. niger, S. mutans, S. aureus, P. aeruginosa,* and *E. coli*	Comparative study of antimicrobial activities of three tropical fruits in Southern India Volume 3 Issue 5 - 2019 TG Madhuri
		Leaves	Acetone, ethyl acetate ethanol, methanol	*S. aureus, S. epidermidis, E. faecalis, E. coli, P. aeruginosa,* and *P. mirabilis*	(Genovese et al., 2020)
7.	Cytotoxic	Root	n-hexane, ethyl acetate, and aqueous acetone	Rat hepatoma cells (dRLh84)	(Hisayoshi et al., 2004)
		Leaves	Aqueous	H9c2	(Wattanapitayakul et al., 2005)
		Leaves, branches, roots, and fruits	methanol	Calu-6, SMU-601, and HCT-116	(Chon et al., 2009)
		Leaves	Aqueous	HepG2	(Naowaratwattana et al., 2010)
		Bark	Methanol	HELA, HEPG2 and MCF7	(Meselhy et al., 2012)
		Leaves	Methanol	HCT-15 and MCF-7	(Deepa et al., 2013)
		Leaves	Aqueous	HepG2	(Fathy et al., 2013)
		Root bark	Methanol	SW480	(Eo et al., 2014)
		Fruits	methanol	MDA-MB-453	(Cho et al., 2017)
		Leaves	ethanol	LL2 cell line	(Fallah et al., 2018)
		Leaves	Ethanol	HeLa cell	(Dabili et al., 2019)

S. no	Activity	Parts used	Extract	Organism used/Cell lines	Reference
8.	Hypotensive, Hypolipidemic, and Vascular protective	Leaves	Aqueous	Male Sprague-Dawley rats	(Lee et al., 2011)
9.	Atherosclerosis	Leaves	Methanol	Human plasma and Male Japanese white rabbit	(Doi et al., 2000)
		Leaves	Aqueous ethanol	LDLR−/− mice	(Enkhmaa et al., 2005)
		Fruits	Aqueous	New Zealand white rabbits	(Chen et al., 2005)
		Leaves	Aqueous ethanol	Venous blood from fasting, healthy adult human	(Katsube et al., 2006)
		Leaves	Aqueous	Wistar rats	(Zeni and Dall'Molin, 2010)
10.	Anti-obesity	Leaves	Ethanol	C57BL/6J mice	(Oh et al., 2009)
11.	Hepatoprotective	Fruits	Aqueous	Male Wistar rats	(Hsu et al., 2012)
12.	Antiplatelet	Leaves	Ethanol	Male Sprague Dawley rats	(D.-S. Kim et al., 2014)
13.	Anxiolytic	Leaves	Methanol	Albino male Swiss mice	(Yadav et al., 2008)
14.	Anthelmintic	leaves	Petroleum ether, chloroform and methanol	*Pheretima posthuma*	(Aditya et al., 2012)
15.	Antioxidant	Leaves	Aqueous	-	(Wattanapitayakul et al., 2005)
		Root bark	Ethanol	-	(El-Beshbishy et al., 2006)
		Fruits	Ethanol	-	(Zhang et al., 2008)
		Leaves, branches, roots and fruits	Methanol	-	(Chon et al., 2009)
		Leaves	Ethanol	-	(Katsube et al., 2009)
		Fruits	Methanol	-	(Imran et al., 2010)
		Leaves, stems and fruits	Aqueous	-	(Wang et al., 2012)
		Leaves, fruits, stem and root bark	Methanolic	-	(Khan et al., 2013)
		Leaves	Water and ethanol	-	(Enmiyet et al., 2014)

Table 11. (Continued)

S. no	Activity	Parts used	Extract	Organism used/Cell lines	Reference
		Leaves	Methanol	-	(D. Kim et al., 2014)
		Fruit	Ethanol	-	(Wang et al., 2013)
		Leaves	Aqueous	-	(Sánchez-Salcedo et al., 2015a)
		Fruits	Methanol	-	(Sánchez-Salcedo et al., 2015b)
		Fruits	Deionized water, aqueous ethanol, aqueous methanol, and aqueous acetone		(Kobus-Cisowska et al., 2020)

Chapter 6

Rhododendron arboreum Sm.: Ethnobotany, Phytochemistry, and Pharmacology

Abstract

Rhododendron arboreum is a medicinal and economically important plant species. It is an evergreen shrub or small tree of the dicotyledonous group which belongs to family Ericaceae. The species is widely distributed from the Western to Eastern Himalayan region and in addition the species is also found growing in other neighboring countries. Different parts of the plant (e.g., leaves, fruits, and roots) possess various pharmacological activities such as antimicrobial, anti-inflammatory, anticancer, hepatoprotective, antidiarrheal, antidiabetic, and antioxidant due to the presence of flavonoids, saponins, tannins, and other phytochemicals. Traditionally, the flowers and leaves have been used for treatment of diarrhea, vomiting, headache, nose bleeding, rheumatism, wound, fever, and blood pressure. Different types of phytochemical components from different parts of the *R. arboretum* have been successfully identified and isolated. The current chapter provides in-depth information about the botanical aspects, traditional uses, phytochemicals and pharmacology of *R. arboretum*.

1. Introduction

India is considered to be the treasure trove of medicinal and aromatic plants. 65 percent of the world's population utilizes herbal medicine for medication, according to the World Health Organization (2000), and 80 percent of the Indian people uses herbal goods for the management of multiple ailments (Kumar et al., 2019). Many species in the mountains grow medicinally and their scientific significance has not been examined. The word Rhododendron derives via the Greek phrase 'rhodo' which means 'rose' and 'dendron' means 'tree' in association with the 'rose tree' (Hora, 1981). The genus *Rhododendron* L. (Family: Ericaceae), is represented by 1025 species (Yu et al., 2017).

Rhododendron is a fairly primitive flowering plant (100 million years) abundant in the temperate parts of the Northern Hemisphere, mainly in the Sino-Himalayas (Eastern Himalayas and Western China). In addition, the *Rhododendrons* have spread to south and northeast China, Japan, Myanmar, Thailand, Malaysia, Indonesia, the Philippines and New Guinea. In India, about 102 species are broadly dispersed in the Himalayas in various regions and altitudes of 1500–5500 m (de Milleville, 2002; Panda and Kirtania, 2016). *Rhododendrons*, the biggest genus in the Ericaceae family make it one of the world's most vulnerable habitats and occupies a larger habitat of the Himalayan subalpine region (Kumar, 2012). In the Himalayas, where major rivers of Asia originate, *Rhododendrons* play an important role in slope stabilization and watershed safety. *Rhododendrons*, however, are the most forgotten scientifically based groups of plants (Kumar, 2012).

R. arborium (de Milleville, 2002) was the first species in India found in 1796 in Kashmir near Srinagar but it was not officially documented until 1817. Major General Thomas Hardwicke discovered *R. arboreum* in 1755-1835 (Lancaster, 1996). Ranging from Sri Lanka and all over Nepal, to Kashmir, to South-East Tibet, *R. arboreum* has a distinctive characteristic of three distinct colors. The pure red flower is evident in the lower height with pink and white flower as its elevation increases.

Rhododendron is a natural plant with many health benefits such as heart disease prevention, diarrhea, detoxification, nausea, headache, constipation, bronchitis, and asthma (Khan et al., 2013). The leaves work successfully as an antioxidant. The young leaves minimize headaches. Khukri handles, saddles, donation boxes, gunstocks and posts can be made with the wood of this plant (Saklani and Chandra, 2015).

R. arboreum is broadly spread from the West to the Eastern Himalayan and other neighboring countries (Giriraj et al., 2008). It is Nepal's national flower and Uttarakhand's state tree. Its flower is recognized as the state flower of Nagaland. The plant is an evergreen shrub/tree disseminated in Bhutan, China, India, Nepal and Pakistan; it is distributed from Kashmir to Nagaland and Western Ghats in the Indian Himalayan region (Orwa et al., 2009). It ranges in height from 4,500 to 10,500 feet (Rai and Rai, 1994) and reaches a height of 12 to 15m but occasionally reaches a height of 30m. The tree was honored the world's largest rhododendron by the Guinness Book of World Records.

The *R. arboreum* is renowned for the ostentatious deep red/crime to pale pink flowers which start blooming from April/June until September. The Indian Postal Service has released a postage stamp that is a tribute to the magnificence of the *R. arboreum* flower. The *R. arboreum* flowers are considered to be holy and is used in temples for decorative purposes. Various parts of the plant have been reported to be of medicinal value (leaves, flowers and root) and are used to treat various ailments in a modern and traditional medicinal method (Swaroop et al., 2005).

Scientific studies have shown that due to the existence of flavonoids, saponins, tannins, and other phytochemicals, the plant is antimicrobial (Nisar et al., 2013; Nisarshy et al., 2013), anti-inflammatory (Verma et al., 2010), hepatoprotective (Verma et al., 2011a), anti-diarrheal, anti-diabetic (Bhandary and Kawabata, 2008) and antioxidant (Acharya et al., 2011, Prakash et al., 2007). Traditionally, the flowers have been used to treat diarrhea, dysentery, headache, nose bleeding, rheumatism, wounds, fever, and blood pressure. The leaf paste is placed on the forehead in headache and fever as an analgesic and antipyretic(Gautam et al., 2016).

2. Botanical Characterization and Classification

2.1. Taxonomical Description

- **Kingdom**: Plantae
- **Phylum**: Magnolliophyta
- **Class**: Angiosperms
- **Order**: Ericales
- **Family**: Ericaceae
- **Genus**: *Rhododendron*
- **Species:** *Rhododendron arboretum*
- **Subspecies**
  - *Rhododendron arboreum* spp. *arboreum* (red or pink red flowers) found in the western Himalayas
  - *Rhododendron arboreum* spp. *cinnamomeun* (white, pink, or red flowers) found in the Central of the Himalayas
  - *Rhododendron arboreum* spp. *delavayii* (red flowers) found in the Eastern Himalayas
  - *Rhododendron arboreum* spp. *nilagiricum* (red flowers) found in Nilgiri
  - *Rhododendron arboreum* spp. *zeylancium* (orange-red flowers) found in Sri Lanka

2.2. Vernacular Names

- **Nepali**: Laligurans
- **Kumaoni**: Eras
- **Punjabi**: Adrawal
- **Tamil**: Billi
- **Kannada**: Pu
- **Malayalam**: Kattuopoovarasu
- **Sanskrit**: Kurvak
- **Garhwali**: Buran
- **Himachal Pradesh**: Brass, Bura

The species is a small evergreen tree as tall as 10m. Leaves clustered at branch ends, leathery, 5-15 cm long, lanceolate to oblong, glabrous dorsal surface, ventral covered with silver scales. Flowers in dense terminal clusters; woolly pedicels; ovate bracts, covered with thick silky hair. Ovate sepals, c. 3 mm long. Corolla crimson, campanulate, reaches 6 cm in length. Stamens 10. 7-9-locular ovaries, intensely woolly. Cylindrical capsule, woolly, 1-2.5 cm long. Seeds many, oblong to ellipsoid, 1.2-3 mm long. Flowering from March - April/June - September bears deep red or crimson to pale pink flowers.

3. Traditional Uses

The source of many scientific discoveries and the production of new information and consumer goods is conventional knowledge. Traditional knowledge of this significance is now generally respected and embraced. The biodiversity range in terms of flora and fauna is limited at high altitudes and hilly areas. Indigenous communities have developed technology and expertise with minimal and locally accessible biological resources in such circumstances to sustain life and livelihoods.

As in many countries, in mountainous regions such as Sikkim, Arunachal Pradesh etc. in India, *Rhododendron* species dominate flora and for decades indigenous populations have gained information about these species and their different uses as part of local traditional knowledge. Many parts of the *R. arboretum* have been reported to have many therapeutical properties and are also used to treat different illnesses as specified in the indigenous medicine system. Its leaves are regularly used in Homeopathic

Materia Medica to treat rheumatism and gout (Rhododendron, 1980). Ashoka Aristha, an ayurvedic formulation that contains *R. arboreum*, is estimated to inhibit estrogenic, oxytocic and prostaglandin synthetase activity. Tender leaves have been documented to be sold as vegetables (Nayyar et al., 1989) However, the use of vegetables is highly questionable as they contain toxic compounds (Andromedotoxin). In cases of elevated temperature and migraine, leaves are used as a poultice.

R. arboreum leaves contain flavonoids, which have a strong antioxidant property (Prakash et al., 2007). *R. arboreum* leaves' ethanolic extract, in normal and isoproterenol-induced myocardial necrosis in rats, has a curative effect on cardiac markers, lipid peroxides and antioxidants. *R. arboreum* leaf extract has powerful analgesic effects based on the dose against peripheral and central analgesic models (Rawat et al., 2017).

The flowers of *R. arboreum* work against blood dysentery, diarrhea and dyspepsia (Laloo et al., 2006; Middelkoop and Labadie, 1983). Occasionally, dried flowers are fried in ghee and then checked for diarrhea (Francis et al., 2007). The flowers of this species were found to be more productive in Shimla hills and other neighboring areas of the Western Himalayas (Biswas, 1956) and may be the climatic factors of the region. The fresh and processed corolla, which is sour to taste, is often used to extract the fish bones that are trapped in the gullet (Pradhan and Lachungpa, 1990). The locals of Himachal Pradesh use their flowers to make pickles (Kiruba et al., 2011). *R. arboreum* flowers are used for making jams, jellies and local brews in mountainous areas. A very popular and fun drink is Rhododendron-wine, popularly called "Guranse" in the Maneybhanjan, Chitrey, Meghma, and Tonglu regions of the Sandakphu-Phallut trek route in the Darjeeling hills. In order to avoid altitude illness, the local brew has been documented to be very effective. Murty et al. (2010) revealed that flowers were registered in experimentally induced hypercholesteric rabbits for the hypolipidemic activity of the ayurvedic herbal formulation Arborium Plus (*Hippophae rhamnoides* and *R. arboreum*). Powdered bark is used as snuff.

In traditional medicine, hot water extract of dried bark is used to treat hypermenorrhea (Middelkoop and Labadie, 1986). The juice and squash not just in Uttarakhand but throughout the country are becoming very famous as a result of their medicinal property. The patient suffering from sugar/diabetics is offered flower juice and squash, used as cardiac tonic, and thought to heal many cardiovascular diseases. The flowers' juice is sometimes used in menstrual disorder therapies (Negi et al., 2013). Taking their root decoction treats cancer early on early stages (Zhasa et al., 2015).

Fresh leaves are burnt with thuja/pine/juniper leaves to create fume which is thought to be sacred and which helps cleanse the air in the region (Paul et al., 2010). Patients with sugar/diabetes who are used as cardiac tonic and believed to cure many heart problems are given bloom juice and squash. (Negi et al., 2013). List of traditional uses of *R. arboreum* are given in Table 12.

4. Chemical Constituent

A number of researchers throughout the globe has conducted phytochemical research on species *R. arboreum* by using various chromatographic extraction techniques. (Rangaswami and Sambamurthy, (1959), Hariharan and Rangaswami (1966), are the researchers who initially carried out phytochemical analysis in this species. The main phytochemicals of the plant include alkaloids, steroids, flavonoids, terpenoids, anthra-quinones, saponins, glycosides and tannins. All these chemicals were isolated from different parts of the plant. Among them, flavonoids and phenols are the main bioactive substances of *R. arboreum* (Kashyap and Zaanand, 2016).

Minerals like manganese, iron, zinc, copper, sodium, chromium, cobalt, cadmium, molybdenum, nickel, lead and arsenic are also present in *Rhododendron*. For the maintenance of certain physical and chemical processes necessary for life, minerals play an important role. Manganese, copper, selenium, zinc, iron and molybdenum are essential cofactors found in the formation of some enzymes and are essential for many biochemical processes. Sodium is essential for maintaining the osmotic balance among cells and interstitial leakage (Soetan et al., 2010).

Table 13 describes the phytochemical ability of various plant parts of the *Rhododendron* species suggesting that the entire plant is a "phytochemical mine," a nutrient-free plant chemical that has protective or immune-modulatory properties (Rangaswami and Sambamurthy, 1960).

5. Pharmacology

As bioactive components are widely distributed throughout the plant kingdom and exhibit a wide variety of biological functions, their phytochemical and pharmacological components have been comprehensively summarized (Agarwal and Kalpana, 1988; Han et al., 2007; Harborne, 2013). A summary of the reported biological activities is given in Table 14.

Table 12. Traditional uses of *R. arboreum*

S.No	Part	Ailment/use	Locality	References
1.	Flowers	Flowers	Uttarakhand, India	(Joshi et al., 2018)
		Honey-mixed drink of flowers is used to cure blurry vision and asthma	Southern China State, Myanmar	(Ong et al., 2018)
		Anti-diabetic, beneficial against diabetic nephropathy (DN), anti-diarrheal activity, antimicrobial activity anti-inflammatory, dyspepsia, anti-nociceptive, Hepatoprotective, fungal infection, free radical scavenging activity, anti-allergy		(Bhattacharyya, 2011; Zha et al., 2016)
		Aqueous extract of flower used as a food coloring agent	Himachal Pradesh, India	(Devi et al., 2018)
		The flowers are eaten raw, sauce and jams are also made	Jammu and Kashmir, India	(Dangwal et al., 2014)
		Dried flowers fried in ghee is used to stop the blood dysentery	Tamil Nadu, India	(Shanmugam et al., 2011)
		The crushed flowers are snuffed to stop the nasal bleeding	Himachal Pradesh, India	(Uniyal et al., 2006)
		The flower and leaves are attached with long ropes made of munja grass and tied around the houses and temples for decoration, women also decorate their hair with flowers	Himachal Pradesh, India	(Kumari et al., 2015)
		Menstrual disorders		(Kumari et al., 2015)
		Flowers juice is sold commercially as a health tonic.	Uttarakhand, India	(Negi et al., 2013)
		Flower juice and squash are used to cure diabetes.		
		Flower juice is used to relieve fever, abdominal pain and heart-related problems.	Uttarkashi, western Himalaya, India	(Nand and Naithani, 2018)
		Flower juice helps when fish bones stuck in the throat.	Sikkim, India	(Tiwari and Chauhan, 2006)
2.	Leaves	Young leaves are applied on the forehead to reduce headache		(Khare, 2008)
		External parasites	Uttaranchal, India	(Pande et al., 2007)

Table 12. (Continued)

S.No	Part	Ailment/use	Locality	References
		Dried leaf tincture is used for the treatment of rheumatism and gout	Homoeopathic materia medica	(Rhododendron, 1980)
		Asoka aristha an Ayurvedic preparation has estrogenic, oxytocic and prostaglandin synthetase inhibition effect		(Middelkoop and Labadie, 1983)
		Antioxidant activity, anti-inflammatory, gout and rheumatics (dried leaves), alleviate headache and fever, diuretics, fungal infection, to relief toothache, sciatica syphilis, treatment of cold cough, asthma, bronchitis, post-delivery complications, indigestion, lung infection, hepatic disorders, analgesic activity, anti-allergy		(Attri et al., 2014; Bhattacharyya, 2011; Kemertelidze et al., 2007)
3.	Bark	Leaves and bark are used to reduce the roughness of the skin	Manoor valley, Pakistan	(Rahman et al., 2018)
		Bark juice is used to cure coughs, piles and liver disorders	Allai valley, Pakistan	(Haq, 2012)
		Alkanoids, steroids, terpenoids, tannins, saponins and reducing sugars. The bark is also found to be a rich source of taraxerol, betulinic acid, ursolic acid ($C_{30}H_{48}O_4$)		(Lepcha et al., 2014; Gill et al., 2015)
4.	Stem Wood	Wood is used to make handles, packsaddles, gift-boxes, Gunstocks and posts	Arunachal Pradesh, India	(Paul et al., 2005)
		Fuel	Arunachal Pradesh, India	(Paul et al., 2010)
		Alkanoids, steroids, terpenoids, anthraquinones, tannins, glycoside saponins and reducing sugar		(Khan et al., 2013)
5.	Roots	The decoction of the roots is used to cure early stage of cancer.	Nagaland, India	(Zhasa et al., 2015)
		Anticancer, anti-nociceptive, anti-infammatory, prevent cardiovascular diseases		(Khan et al., 2013; Srivastava, 2012)

Table 13. Phytochemicals present in *Rhododendron arboreum* plants

S. No.	Part of plant	Compound	References
1.	Leaves	Quercetin 3-O- beta -Dglucopyranosyl [1-6]-O-alpha-L-rhamnopyranoside	(Kamil and Ilyas, 1995)
		2-stigmasten-3-one, 1,1,6-trimethyl-3-methylene-2-(3,6,10,13,14-pentamethyl-3-ethenyl-pentadec-4-enye) cyclohexane, alpha-amyrin, beta-amyrin, linoleyl alcohol, linoleic, beta-citronellol, tetradecane, 9,12-octadecadienoic acid, methyl ester, dibutyl phthalate, L-ascorbic acid 2, 6-dihexadecanoate, dodecane, and heptadecane	(Pahnuli et al., 2016)
		3,7,11,15-tetramethyl-2-hexadecen-1-ol, linoleic acid, 9-octadecenoic acid, 3,10-epoxy-(3β,10β)-D:B-friedo-18,19-secolup-19-ene, methyl commate C, 2-Hexyl-1-decanol, OctadecylChloroacetate, 3-bromo-3β-cholest-5-ene, Methyl commate B andolean-12-en-28-al, 1-Tetradecene, Eophytadiene, 3(E)-eicosene and 9(E)-eicosene, 2,6,10,15,19,23-Hexamethyl-2,6,10,14,18,22-tetracosahexaene, and vitamin E	(Gautam et al., 2018)
		Quercetin 3-O- beta -D-glucopyranosyl 1, 2'-dimethoxy-4',5-methylene dioxyflavanone, L-rhamnopyranoside, Pectolinarigenn 7 orutinoside	(Kamil and Ilyas, 1995)
		pectolinarigenin 7-Orutinoside, 7, 2'-dimethoxy-4', 5-methylenedioxyflavanone	(Kamil and Ilyas, 1995)
		Glucoside, Ericolin, Ursolic acid, α-amyrin, Epifriedelinol, Campanulin, Quercetin, and Hyperoside	(Orwa et al., 2009)
		Quercetin-3-o-galactoside, Epicatechin, Quercitrin and Syringic acid	(Sharma et al., 2010)
		Flavonoid, Tannin, Steroid and Terpenoid in flower; Flavonoid, Anthraquinones, Tannins, Steroids and Terpenoids	(Nisar et al., 2016)
		Phenols, Flavonoids, Gallic acid, Quercetin	(Roy et al., 2014)
		Ursolic acid, β-sitosterol, Lupeol	(Shilajan and Swar, 2013)
		Pyrogoll and Catechol	(Saklani and Chandra, 2015)

Table 13. (Continued)

S. No.	Part of plant	Compound	References
		Glucoside, Ericolin, Ursolic acid, Quercetin, Hyperoside, Flavone glycosides, Flavonoids	(Verma et al., 2011a)
2.	Flowers	Rutin, Quercetin and Coumaric acid	(Sonar et al., 2012b; Swaroop et al., 2005)
		Quercetin-3-O-rhamnoside	(Rangaswami and Sambamurthy, 1959)
		Pyrogoll and catechol	(Saklani and Chandra, 2015)
		Saponins, Proteins, Steroids, Tannins, Xanthoproteins, Coumarins and Carbohydrates	(Kiruba et al., 2011)
		Hyperin	(Verma et al., 2011a)
3.	Stem	Alkaloids, Tannins, Saponin reducing sugars, Terpenoids and Steroids	(Nisar et al., 2016)
4.	Bark	Alkaloids, Tannins, Saponin reducing sugars, Terpenoids and Steroids	(Nisar et al., 2016)
		Taraxerol, Betulinic acid, Leuco-pelargonidin and Ursolic acid acetate	(Hariharan and Rangaswami, 1966)
		Pyrogoll and Catechol	(Saklani and Chandra, 2015)
		Betulinic acid	(Rawat et al., 2017)
		4,4,6a,6b,11,12,14b–Heptamethyl-16-oxo1,2,3,4,4a,5,6,6a,6b,7,8,9,10,11,12,-12a,14a,14b–octadecahydro -12b,8a-(epoxymethano) picen-3-yl acetate	(Khan et al., 2013)
5.	Roots	Alkaloids, Tannins, Saponin reducing sugars, Terpenoids and Steroids	(Nisar et al., 2016)

Table 14. Pharmacological activities of *R. arboreum*

S. No	Activity	Parts used	Extract	Organism used	Reference
1.	Antibacterial	Flowers	Ethanolic and Aqueous	*B. subtilis*, *E. coli*, *S. aureus*, *aeruginosa*, *A. tumefaciens*, *C. albicans*, and *A. niger*	(Sonar et al., 2012b)
		Leaves, roots, flowers, stem, and bark	Methanolic	*Bacillus subtills*, *Staphylococcus aureus*, *Salmonella typhi* and *Escherichia coli*	(Nisarshy et al., 2013)
		Whole plant	Ethanolic	*Bacillus subtilis*, *Staphylococcus aureus*, *Escherichia coli* and *Salmonella typhi*	(Sharma, 2013)
		Leaves	Petroleum ether and chloroform	*Staphylococcus aureus*, *Bacillus subtilis*, *Proteus vulgaris*, and *E. coli*.	(Jegan and Selvaraj, 2014)
		Leaves and flowers	Alcoholic	*S. aureus*	(Chauhan et al., 2016)
		Leaves	Methanolic	*Escherichia coli*, *Yersinia pestis*, *Bacillus cereus*, *Pseudomonas aeruginosa*, *Listeria monocytogenes*, and *Staphylococcus aureus*	(Prakash et al., 2016)
		Flowers	Ethanolic	*Bacillus Subtilis*, *Bacillus cereus*, *Staphylococcus aureus*, and *Escherichia coli*, *Salmonella enteric* and *Shigella flexineri*	(Kashyap et al., 2017)
		Flower	Ethanolic, methanolic, and acetone	*B. cereus*, *E. coli*, *Salmonella typhi*, *S. aureus*, *Shigellaflexneri*, and *P. aeruginosa*	(LAL et al., 2017)
		Bark	Aqueous and methanol	*Streptococcus pyogene*, *Staphylococcus aureus*, *Escherichia coli*, and *Klebsiella pneumonia*,	(Saranya and Ravi, 2017)

Table 14. (Continued)

S. No	Activity	Parts used	Extract	Organism used	Reference
		Flower	Aqueous and methanol	Streptococcus pyogene, Staphylococcus aureus, Escherichia coli, and Klebsiella pneumonia,	(Saranya and Ravi, 2017)
2.	Antifungal	Bark	n-hexane, chloroform, ethyl acetate, and methanol	Aspergillus niger, Candida albicans, Cantharellus flavus, Fusarium solani, Microsporum canis and Diogma glabrata	(Khan et al., 2013)
		Leaves	Ethanolic	Proteus vulgaris and S. Aureus	(Jegan and Selvaraj, 2014)
		Flower	Ethanolic	Aspergillus flavus, Candida albicans, Aspergillus parasiticus, Candida albicans, Aspergillus Parasiticus, and Aspergillus flavus	(Saklani and Chandra, 2015)
		Leaves	Methanolic	Trichoderma viride	(Chauhan et al., 2016)
		Leaves	Aqueous, methanolic	Streptococcus pyogene, Staphylococcus aureus, Escherichia coli, Klebsiella pneumonia, Candida albicans, and Trichoderma viride	(Saranya and Ravi, 2016)
		Bark	Aqueous and methanol	Candida albicans and Trichoderma viride	(Saranya and Ravi, 2017)
		Flower	Aqueous and methanol	Candida albicans and Trichoderma viride	(Saranya and Ravi, 2017)
3.	Cytotoxic	Flowers and	Ethanolic	MCF-7	(Sonar et al.,

S. No	Activity	Parts used	Extract	Organism used	Reference
		leaves			2012a)
		Leaves, roots, flowers, stem, and bark	Methanolic	*Artemia salina*	(Nisarshy et al., 2013)
		Leaves	Aqueous	-	(Painuli et al., 2018)
		Leaves and flowers	Methanolic	HeLa, MCF7, and A549	(Gautam et al., 2020)
4.	Anti-inflammatory and Anti-nociceptive	Flowers	Ethyl acetate	Wistar albino rats and Swiss albino mice	(Verma et al., 2010)
		Bark	Methanolic	Male BALB/c mice	(Nisar et al., 2016)
5.	Anti-inflammatory	Flowers	Aqueous, 50% ethanolic and methanolic	-	(Agarwal and Kalpana, 1988)
		Leaves	Ethanolic	Wistar albino rats and Swiss albino mice	(Sharma et al., 2009)
6.	Antimutagenic activity	Leaves	Hexane, chloroform, and ethyl acetate	*S. typhimurium* (TA-98 and TA-100)	(Gautam et al., 2018)
7.	Antidiarrheal	Flowers	n-hexane, chloroform, ethyl acetate and n-butanol	Wistar albino rats and Swiss albino mice	(Verma et al., 2011b)
8.	Antidiabetic activity and Anti-hyperlipidemic	Bark	Ethanolic	Albino Wistar male rats	(Gautam and Chaudhary, 2020)

Table 14. (Continued)

S. No	Activity	Parts used	Extract	Organism used	Reference
9.	Antidiabetic	Flower	Methanolic	Rats	(Bhandary and Kawabata, 2008)
10.	Antihyperglycemic and Antihyperlipidemic	Flowers	Ethanolic	Male albino Wistar rats	(Verma et al., 2012)
11.	Antihyperglycemic	Flowers	n-hexane, chloroform, ethyl acetate and n-butanol	Male Wistar rats	(Verma et al., 2013)
12.	Cardioprotective activity	Whole plant	Ethanolic	Male albino Wistar Rat	(Mudagal and Goli, 2011)
13.	Immunomodulatory	Leaves	Ethanolic	Swiss albino mice	(Sonar et al., 2013)
14.	Hepatoprotective	Leaves	Ethanolic	Wister Albino rats, and Albino mice	(Prakash et al., 2008)
15.	Myocardial infarction	Leaves	Ethanolic	Male albino Wistar rats	(Mudagal and Goli, 2011)
16.	α-glucosidase inhibitor	Bark	Methanolic	-	(Raza et al., 2015)

Conclusion

Rhododendron arboreum is a medicinal plant that has been used by tribals for a long time. The ethnomedical, phytochemical and pharmacological implications of this medicinal plant are concisely summarized in this chapter. Due to the different medicinal value of *R. arboreum*, numerous studies have been conducted on the purification of bioactive components in the *R. arboreum* and their analysis in terms of chemical nature and biopharmacological function. The productivity of crude extracts and refined ingredients in disease therapy created opportunities for research and development on drugs.

Chapter 7

Shorea robusta Gaertn. f.: Ethnobotany, Phytochemistry, and Pharmacology

Abstract

Shorea robusta is considered as an important medicinal tree which is used in Indian traditional medicine for the treatment of various diseases such as digestive, respiratory, circulatory as well as for infectious diseases. The phytochemical analysis in the species have revealed the occurrence of several secondary metabolites belonging to terpenoids, flavonoids, sterols and phenols. The crude extracts and isolated chemicals from the species have shown a wide range of biological activities such as anti-inflammatory, anti-fungal, anti-bacterial, anti-obesity, anti-pyretic and wound healing activities. The current chapter offers comprehensive information on botany, traditional uses, phytochemistry and pharmacology of *S. robusta*.

1. Introduction

Sal (*Shorea robusta* Gaertn. f.) is a perennial tree species of the family Dipterocarpaceae. It is a huge deciduous tree, approximately 18-32m in height and 1.5-2m in circumference. The bark of the plant is dark brown and dense, with deep longitudinal cracks in poles. Leaves are simple, glossy and glabrous, about 10-25 cm long and broadly oval at the base with the long-pointed apex. The flowers are arranged in large terminal or axillary racemose panicles, which are yellowish-white. The fruit is about 1.3-1.5cm long and 1cm in diameter when fully mature.

The species is widely distributed in three Himalayan countries viz., Bhutan, Nepal and India. In India, the species is found growing in Himachal Pradesh, Tripura, West Bengal, Bihar, Orissa, Madhya Pradesh and Andhra Pradesh (Fu, 1994; Zhao et al., 1994). The species is found mainly on the plains and lower foothills of the Himalayas (Gautam, 1990). The natural range of the species lies within longitudes of 75° and 95° E and latitudes of

20° to 32° N and within this range, the species distribution is controlled mainly by climate and edaphic factors.

In the past, sal forests were managed mainly due to the interests of elite people and its management rules were established accordingly to increase revenue (Gadgil, 1990, Gadgil and Guha, 1993), but now a day, the government is managing sal forests for commercial timber production as well as other products to order to increase its revenue. Sal forests have the ability to provide extra forest resources other than timber as well. Along with wood and fuel, a sal tree produces fodder (Edwards, 1996; Fox, 1995; Gautam, 1990; Gautam and Devkota, 1999; Kk, 1982; Mathema, 1991; Pandey and Yadama, 1990; Shakya and Bhattarai, 1995; Thacker and Gautam, 1994; Upadhyay, 1992); leaves for making plates (Gautam and Devkota, 1999; Rajan, 1995); oil seed (Sharma, 1981; Venkat and Sharma, 1978); feed (Sinha and Nath, 1982), resin or latex from heartwood and tannin and gum from bark (Narayanamurti and Das, 1951; Karnik and Sharma, 1968).

The species associated with sal are known to produce grow edible fruits, forage and manure, fibers, leaves for umbrellas, medicinal plants, hay, brooms and many other things depending on the composition of the species (Jolly et al., 1976; Panday, 1982; Gautam, 1990; Amatya, 1990; Gilmour and Fisher, 1991; Suresh, 1994; Jackson et al., 1994; Fox, 1995; Thacker and Gautam, 1994; Shakya and Bhattarai, 1995; Tewari, 1995; Edward, 1995; Sah, 1996; Stainton, 1972; APROSC, 1995; Dwivedi, 1993; Melkania and Ramnarayan, 1998; Poudyal, 2000; Webb and Sah, 2003). Besides, there are appealing aspects about the traditional methods of lopping, browsing and litter gathering in the sal forests of Nepal and surrounding areas (Agrawal et al., 1986; Bahuguna, n.d.; Banerjee and Mishra, 1996; Bhat and Rawat, 1995; Chopra and Chatterjee, 1990; Dinerstein, 1979; Melkania and Ramnarayan, 1998; Mukhopadhyay, 1991; Nepal and Weber, 1995; Pandey and Yadama, 1990; Ram Prasad and Pandey, 1987; Rao and Singh, 1996; Saxena et al., 1993; Sundriyal et al., 1994). The proof of such a wide variety of sal forest products suggests that with proper management, many associated sal forest species are capable of generating useful products.

Sal prefers to grow in different types of soils, the species can grow in alluvial and lateritic soils (Tewari, 1995) having an organic carbon content around 0.11 and 1.8 percent (Gangopadhyay et al., 1990; Rama et al., 1988). Sal forests extend into tropical and subtropical areas as well as to regions where precipitation varies from 1000-2000 mm with no more than four months of dry weather (Tewari, 1995). Sal tolerates some frost, but seedlings are destroyed by extreme annual frost occurring in frost hole (Prasad and Pandey, 1987). In the sal forest the highest temperature recorded is 49°C (Singh et al., 1993; Singh and Chaturvedi, 1983).

The phenology of a sal forest varies from deciduous to evergreen and extends from tropical to subtropical based on edaphic variables and microclimate. In late winters (i.e., in February), the leaf fall typically starts and ends in late April (Misra, 1969). Since the forest comprises of numerous species, particularly *Butea monosperma*, *Cadrela toona*, *Mangifera indica*, *Syzgium cumini*, *Terminalia tomentosa*, the phenology of sal correlates with the phenology of these species. Maximum leaf fall is observed from mid-February to mid-May (Pokhriyal et al., 1987; Singh et al., 1993). Sal trees yield seeds every year and the best seed set is seen every third year. Sal seed production ranges from year to year and from tree to tree (Tewari, 1995).

Germination of seedling is usually reported to occur from mid-May to mid-June. Sal seeds have wings and are scattered in the air approximately 100 m from the mother tree (Jackson et al., 1994). The rate of germination is high (over 90 percent), provided there is enough rain available for a seed within a week. Large number of seeds are reported to germinate yearly. The seed loses its viability in one week, so the seed will no longer germinate if the monsoon that typically begins at the end of June is delayed.

Being a light-loving species, it needs bright light right from the initial stages of its growth (Champion and Seth, 1968; Kayastha, 1985). Regeneration and the development of understory seedlings and saplings are stimulated by the opening of the canopy in a forest stand (Gautam, 1990). But certain side shades could be beneficial in dry environments, and juvenile plants may need shelter from frost and drought (Jackson, 1994; Tewari, 1995).

1.1. Taxonomical Description

- **Kingdom**: Plantae
- **Subkingdom**: Viridaeplantae
- **Infrakingdom**: Streptophyta
- **Division**: Tracheophyta
- **Subdivision**: Spermatophytina
- **Infrared division**: Angiospermae
- **Class**: Magnoliopsida

- **Superorder**: Rosanae
- **Order**: Malvales
- **Family**: Dipterocarpaceae
- **Genus:** *Shorea*
- **Species**: *Shorea robusta* C.F. Garten

1.2. Vernacular Names

- **Arabic:** Qanquahar
- **Assamese:** Sal-goch, Sal, Hal, Borsal
- **Ayurveda:** Shala
- **Bengali:** Sakhu, Sal, Shal, Sakher, Salwa
- **Burmese:** Suo Luo Shuang, Suo Luo Shuang Shu, Enkhyen
- **Chinese:** Suo Luo Shuang, Suo Luo Shuang Shu
- **English:** Sal trees, Common Sal, Indian Dammer, Yellow Balau.
- **French:** French: Damar de l'Inde, Arbre à Sal, Balau Jaune.
- **Gujarati:** Rala, Khakhra Tree, Ral
- **Hindi:** Borsal, Hal, sakhu, Sakhwa, Sal, Shal, Damar, Dhuna, Salwa
- **Indian:** Jall, Sal, Salwa, Shal
- **Japanese:** Bangkirai (Borneo), Damar Laot (Indonesia), Selangan Batu (Sabah), Selangan Batu Kumus, Sara Noki, Serangan Batsuu, Shara Noki
- **Marathi:** Guggilu, Rala, Sal, Sajara.
- **Punjabi:** Sal, Seral
- **Russian:** Sal, Salovoe Derevo, Shoreia Moshchania.
- **Sanskrit:** Ashvakarna, Sheraparna, Sal, Sala, Sarga, Guggilam, Sasyasambara, Tarkshyaprasava, Dhupavrksa, Sarjakarya, Koushikaha, Ashvakarnika, Agnivallabha
- **German:** Salbaum, Salharzbaum
- **Nepali:** Agrakh, Sakhua, Sakwa, Sal
- **Tamil:** Kungiligam, Attam, Shalam, Saruva rasam, Kukkil, Acuvakarnakam, Kunkiliyamaram, Salam, Nitiyopam, Attam, Ulukalam, Pantitturumaram, Sulukacamaram, Venkungiliyam, Kungiliyam
- **Telugu:** Gugal, Guggilamu, Saluva, Sarjmu
- **Urdu:** Safed Dammar, Ral, Safad dammar

2. Traditional Uses of *Shorea robusta*

S. robusta has been traditionally used for curing a number of diseases. The leaves and bark are used for curing cuts, ulcers, leprosy, cough, gonorrhea, earache, and headache (Duddukuri et al., 2011; Jyothi et al., 2008). The oleoresin derived from the cut bark has astringent, carminative and stomachic properties (Kalaiselvan et al., 2013). It is useful in treating diarrhea, cuts, ulcers, neuralgia, burns, bruises, fever, dysentery, splenomegaly, obesity, and eye burning (Murthy, 2011; Kalaiselvan and Gokulakrishnan, 2012; Mathavi and Nethaji, 2014; Mukherjee et al., 2013).

The fruits are helpful for tubercular ulcers, seminal fatigue, burning sensation, and dermopathy. In Unani medication, the resin is used for menorrhagia treatment (Pullaiah and Rani, 1999), enlargement of the spleen (Misra and Ahmad, 1997), and for reducing eye irritation (Patra et al., 1992). In Ayurveda, it is used with sugar or honey in treating dysentery and bleeding piles. It is also recommended for gonorrhea and poor digestion. It has been recommended by Siddha practitioners for ulcers, injuries, and postmenopausal disorders (Dey et al., 2012; Kaur et al., 2003; Shafiuddin et al., 2009; T A Wani et al., 2012a).

S. robusta leaf extract has been discovered to have strong anti-inflammatory action (Manoj et al., 2014). It has been shown that a mixture of cow ghee, flaxseed oil, *Phyllanthus emblica* fruits, *S. robusta* resin, and *Yashada bhasma* has wound healing activity (Mukherjee et al., 2013).

3. Phytoconstituents

The wide varieties of phytoconstituents have been found from various parts of *S. robusta* by a number of researchers from different parts of the globe. Table 15 lists extracted compounds from the different parts of the plant.

4. Pharmacological Properties

Various studies have been carried out so far and published on the pharmacological activities of *S. robusta* and these studies have revealed that the different parts of this plant species exhibit various biological properties including anti-inflammatory, anti-fungal, anti-bacterial, antioxidant and analgesic. These biological properties are due to isolated phytochemicals which can be of great interest for the development of plant-based medicine. The detailed pharmacological activities are given in Table 16.

Table 15. Classes and names of compounds isolated from the various parts of *S. robusta*

Sr. no.	Compound	Part of plant	References
1.	Ursolic acid	Whole plant	(Hota and Bapuji, 1993; Misra and Ahmad, 1997)
2.	α-amyrenone		
3.	α-amyrin		
4.	β-amyrin	Leaves & whole plant	(Chauhan et al., 2002; Hota and Bapuji, 1993; Misra and Ahmad, 1997)
5.	Shoreaphenol	Bark	(Patra et al., 1992)
6.	Hopeaphenol	Seed	(Prakash and Rao, 1999)
7.	Friedelin, β-sitosterol, α- and β-carotene, lutein, phenophytin, 7-methoxy-4'-5-dihydroxyisoflavone	Leaves	(Chauhan et al., 2002)
8.			
9.	Dihydroxyisoflavone		
10.	Asiatic acid, 3,25-epoxy-1,2,3,11-tetrahydroxyurs-12-en-28-oic Acid, 3,25-epoxy-1,2,3-trihydroxyurs-12-en-28-oic Acid, Phayomphenol and 3,7-dihydroxy-8-methoxyflavone 7-O-α-L-rhamnopyranosyl-(1→4)-α-L-rhamnopyranosyl-(1→6)-βD-glucopyranoside.	Root Bark	(Sharma et al., 2014)
11.	3,7-dihydroxy-8-methoxyflavone 7-O-α-l-rhamnopyranosyl-(1→4)-α-l-rhamnopyranosyl-(1→6)-β-d-glucopyranoside	Seeds	(Prakash and Rao, 1999)
12.	2α,3β,23-trihydroxy-11β- methoxy-urs-12-en-28-oic acid	Resin	(Hota and Bapuji, 1994)
13.	Trimethylsilyl 3-methyl-4-[(trimethylsilyl)oxy] benzoate, D-Mannitol, Sorbital, Phytol, Hexamethylcyclotrisiloxane, β-Caryophyllene, 1,2,4-Benzenetriol	Bark, leaf, flower, and seed	(Marandi et al., 2016)
14.	Benzofuran, Shoreaphenol	Bark	(Patra et al., 1992)

Table 16. An overview of the modern pharmacological assessments of *S. robusta*

S. no	Activity	Parts used	Extract	Organism used/Cell lines	Reference
1.	Anti-bacterial	Bark, leaves, flower, and seed	Ethanolic	*Bacillus cereus*, *B. subtilis*, *Enterobacter aerogenes*, *Escherichia coli*, *Klebsiella pneumonia*, *Proteus mirabilis*, *P. vulgaris*, *Pseudomonas aeruginosa*, *Salmonella paratyphi*, *Staphylococcus aureus*, *S. faecalis*, *Vibrio cholerae*	(Marandi et al., 2016)
		Resin	Methanol	*S. aureus*, *B. subtilis*, *P. aeruginosa*, *E. coli*	(Vashisht et al., 2016)
		Flowers	Aqueous	*S. aureus*, *B. subtilis*, *K. pneumonia*, *Serratia marcescens*	(Duddukuri et al., 2011)
		Leaves		*S. aureus*, *K. pneumonia*	(Archana and Jeyamanikandan, 2015)
		Gum	Petroleum ether, methanol, benzene, and aqueous	*Bacillus subtilis* (G+) NCIM 2063 *Bacillus licheniformis* (G+) MTCC 1483 *Bacillus coagulans* (G+) NCIM 2030 *Bacillus cereus* (G+) MTCC 1305 *Salmonella typhi* (G+) MTCC 537 *Staphylococcus aureus* (G+) NCIM 2079 *Staphylococcus epidermidis* (G+) NCIM 2493 *Staphylococcus griseus* (G+) MTCC 1540 *Escherichia coli* (G-) NCIM 2065 *Proteus vulgaris* (G-) NCIM 2857 *Pseudomonas fluoresence* (G-) MTCC 748 *Aspergillus flavus* MTCC 1884 *Aspergillus niger* MTCC 1785 *Candida albicans* MTCC 1637 *Penicillium chrysogenum* MTCC 1996	(Murthy, 2011)

Table 16. (Continued)

S. no	Activity	Parts used	Extract	Organism used/Cell lines	Reference
2.	Immuno-modulatory	Bark	Ethanol	Male Swiss albino mice	(Kalaiselvan and Gokulakrishnan, 2012)
3.	Anti-obesity	Leaves	Hydro-alcoholic	Male albino rats	(Supriya et al., 2012)
4.	Anti-inflammatory	Leaves	Methanol	Swiss albino male mice	(Debprasad et al., 2012)
				Wistar rats and albino mice	(Jyothi et al., 2008)
		Resin	Ethanol	Male Wistar albino rats	(T A Wani et al., 2012b)
5.	Analgesic	Leaves	Methanol	Swiss albino male mice	(Jyothi et al., 2008)
		Resin	Ethanol	Wistar albino rats	(Tariq Ahmad Wani et al., 2012)
6.	Antipyretic	Resin	Ethanol	Male Wistar albino rats	(T A Wani et al., 2012b)
7.	Antioxidant	Resin	Methanol	-	(Vashisht et al., 2016)
		Leaves	Methanol	-	(Archana and Jeyamanikandan, 2015)
		Bark	Ethanol	Male Wistar albino rats	(Kalaiselvan et al., 2013)
8.	Antinociceptive	Leaves	Methanol	Wistar rats and albino mice	(Jyothi et al., 2008)
9.	α- amylase inhibition	Leaves	Methanol	-	(Archana and Jeyamanikandan, 2015)
10.	Anti-obesity	Leaves	Hydro-alcoholic	Male albino rats	(Supriya et al., 2012)
11.	Antiulcer	Resin	Coconut water	Male Wistar albino rats	(Santhoshkumar et al., 2012)
12.	Wound healing	Resin	Methanol	Male Wistar albino rats	(Yaseen Khan et al., 2016)
		Resin	Ethanol	Male Wistar albino rats	(T A Wani et al., 2012a)

Conclusion

The scientific work published on *S. robusta* proved it as an important medicinal plant for a number of medical procedures. The plant has been in operation for a long time and has no noticeable adverse impact. The comprehensive knowledge provided in this chapter shows support for its phytochemical, pharmacological & traditional applications. The results of these future works will supply resources of phytochemicals with great potential for the pharmaceutical industry.

References

Abbasi A.M, Khan M.A, Ahmad, M, Zafar, M, Khan, H, Muhammad N, Sultana S. 2009. "Medicinal plants used for the treatment of jaundice and hepatitis based on socio-economic documentation". *African Journal of Biotechnology* 8.

Abbasi AM, Khan MA, Ahmed M, Zafar M. (2010). Herbal medicines used to cure ailments by inhabitants of Abbottabad district, North West Frontier Province, Pakistan. *Indian J Trad Know* 9:175–83.

Acharya, K., Giri, S., Biswas, G., 2011. Comparative study of antioxidant activity and nitric oxide synthase activation property of different extracts from *Rhododendron arboreum* flower. *Int J Pharmtech Res* 3, 757–762.

Aditya R.S.J, Ramesh C.K, Riaz M, Prabhakar B.T. 2012. Anthelmintic and antimicrobial activities in some species of mulberry. *International Journal of Pharmacy and Pharmaceutical Sciences* 4: 335–338.

Afzal, S., Afzal, N., Awan, M.R., Khan, T.S., Gilani, A., Khanum, R. and Tariq, S., 2009. Ethno-botanical studies from Northern Pakistan. *J Ayub Med Coll Abbottabad*, 21(1), pp.52-7.

Agarwal, S.S., Kalpana, S., 1988. Anti-inflammatory activity of flowers of *Rhododendron arboreum* (Smith) in Rat's hind paw oedema induced by various phlogistic agents. *Indian J. Pharmacol.* 20, 86.

Aghel, N., Kalantari, H. and Rezazadeh, S., 2011. Hepatoprotective effect of *Ficus carica* leaf extract on mice intoxicated with carbon tetrachloride. *Iranian Journal of Pharmaceutical Research: IJPR*, 10(1), p.63.

Agrawal, A.K., Joshi, A.P., Kandwal, S.K., Dhasmana, R., 1986. An ecological analysis of Malin riverain forest of outer Garhwal Himalaya(Western Himalaya). *Indian J. Ecol.* 13, 15–21.

Ahmad, J., Khan, S. and Iqbal, D., 2013. Evaluation of antioxidant and antimicrobial activity of *Ficus carica* leaves: an in vitro approach. *Journal of Plant Pathology & Microbiology.*

Ahmad, R., Srivastava, S.P., Maurya, R., Rajendran, S.M., Arya, K.R. and Srivastava, A.K., 2008. Mild antihyperglycaemic activity in *Eclipta alba, Berberis aristata, Betula utilis, Cedrus deodara, Myristica fragrans* and *Terminalia chebula*. *Indian Journal of Science and Technology*, 1, 1–6.

Ahmad, S., Bhatti, F.R., Khaliq, F.H., Irshad, S. and Madni, A., 2013. A review on the prosperous phytochemical and pharmacological effects of *Ficus carica*. *International Journal of Bioassays*, 2, pp.843-849.

Akbarzadeh, T., Sabourian, R., Saeedi, M., Rezaeizadeh, H., Khanavi, M., Ardekani, M.R.S., 2015. Liver tonics: review of plants used in Iranian traditional medicine. *Asian Pac. J. Trop. Biomed.* 5, 170–181.

Al Yousuf, H.H.H., 2012. Antibacterial activity of *Ficus carica* L. extract against six bacterial strains. *International Journal of Drug Development and Research* 2012, Volume 4, Number 4; Page(s) 307 To 310.

Alam, M.K., 1995. Diversity in the woody flora of sal (*Shorea robusta*) forests of Bangladesh. *Bangladesh J. For. Sci.* 24, 41–51.

Alasalvar, C., Amaral, J.S., Shahidi, F., 2006a. Functional lipid characteristics of Turkish Tombul hazelnut (*Corylus avellana* L.). *J. Agric. Food Chem.* 54, 10177–10183.

Alasalvar, C., Bolling, B.W., 2015. Review of nut phytochemicals, fat-soluble bioactives, antioxidant components and health effects. *Br. J. Nutr.* 113, S68–S78.

Alasalvar, C., Karamać, M., Amarowicz, R., Shahidi, F., 2006b. Antioxidant and antiradical activities in extracts of hazelnut kernel (*Corylus avellana* L.) and hazelnut green leafy cover. *J. Agric. Food Chem.* 54, 4826–4832.

Alasalvar, C., Shahidi, F., 2008. *Tree Nuts: Composition, Phytochemicals, and Health Effects*. CRC Press.

Ali, B., Mujeeb, M., Aeri, V., Mir, S.R., Faiyazuddin, M. and Shakeel, F., 2012. Anti-inflammatory and antioxidant activity of *Ficus carica* Linn. leaves. *Natural Product Research*, 26(5), pp.460-465.

Amaral, J.S., Ferreres, F., Andrade, P.B., Valentão, P., Pinheiro, C., Santos, A., Seabra, R., 2005. Phenolic profile of hazelnut (*Corylus avellana* L.) leaves cultivars grown in Portugal. *Nat. Prod. Res.* 19, 157–163.

Amaral, J.S., Valentão, P., Andrade, P.B., Martins, R.C., Seabra, R.M., 2010. Phenolic composition of hazelnut leaves: Influence of cultivar, geographical origin and ripening stage. *Sci. Hortic.* (Amsterdam). 126, 306–313.

Amarowicz, R., Dykes, G.A., Pegg, R.B., 2008. Antibacterial activity of tannin constituents from *Phaseolus vulgaris, Fagoypyrum esculentum, Corylus avellana* and *Juglans nigra*. *Fitoterapia* 79, 217–219.

Amatya, S.M., 1990. *Fodder trees and their lopping cycle in Nepal.*

Amol, P.P., Vikas, V.P., Vijay, R.P. and Rajesh, Y.C., 2010. Anthelmintic and preliminary phytochemical screening of leaves of *Ficus carica* Linn against intestinal helminthiasis. *International Journal of Research in Ayurveda and Pharmacy (IJRAP)*, 1(2), pp.601-605.

Ara, A., Naqvi, S.H., Rehman, N.U. and Qureshi, M. M. 2020. Comparative antioxidative and antidiabetic activities of *Ficus carica* pulp, peel and leaf and their correlation with phytochemical contents. *Pharm Res,* 4(2).

Arabshahi-Delouee S, Urooj A. 2007. Antioxidant properties of various solvent extracts of mulberry (*Morus indica* L.) leaves. *Food Chemistry* 102: 1233–1240.

Archana, K., Jeyamanikandan, V., 2015. In vitro antibacterial, antioxidant and α-amylase inhibition activity of medicinal plants. *J. Chem. Pharm. Res.* 7, 1634–1639.

Aref, H.L., Salah, K.B., Chaumont, J.P., Fekih, A., Aouni, M. and Said, K., 2010. In vitro antimicrobial activity of four *Ficus carica* latex fractions against resistant human pathogens (antimicrobial activity of *Ficus carica* latex). *Pak J Pharm Sci,* 23(1), pp.53-58.

Aryal, B., Giri, A., Shrestha, K.K., Ghimire, S.K., Jha, P.K., 1999. Vegetation analysis of *Shorea robusta* forests in the Royal bardia National Park, Nepal. *Bangladesh J. Bot.* 28, 35–46.

Asano N, Oseki K, Tomioka E, Kizu H, Matsui K. 1994. N-containing sugars from *Morus alba* and their glycosidase inhibitory activities. *Carbohydrate Research* 259, 243–255.

Attri, B.L., Krishna, H., Ahmed, N., Kumar, A., 2014. Effect of blending and storage on the physico-chemical, antioxidants and sensory quality of different squashes. *Indian J. Hortic.* 71, 546–553.

Awasthi, A.K., Nagaraja, G.M., Naik, G. V, Kanginakudru, S., Thangavelu, K., Nagaraju, J., 2004. Genetic diversity and relationships in mulberry (genus Morus) as revealed by RAPD and ISSR marker assays. *BMC Genetics* 5: 1–9

Ay, E. and Duran, N., 2018. Investigation of the antiviral activity of *Ficus carica* L. latex against HSV-2. In International Conference on Advanced Materials and Systems (ICAMS) (pp. 33-37). *The National Research & Development Institute for Textiles and Leather-INCDTP.*

Azam FA, Qadir MI, Uzair M, Ahmad B. 2013. Evaluation of antibacterial and antifungal activity of *Ficus carica* and *Cassia fistula*. *Lat. Am. J. Pharm.* 2013 Jan 1;32(9): 1412-4.

Azam, F., Rehman, S. Ur., Hamad, H.A., Khalid, A., Samad, A. 2019. In-vitro assessment of anti-bacterial, anti-fungal and anti- oxidant potential of *Ficus carica* stems bark. *International Journal of Biosciences.* Vol. 14, No. 1, p. 520-524.

Aziz, F.M., 2012. Protective Effects of Latex of *Ficus carica* L. against Lead Acetate-Induced Hepatotoxicity in Rats. *Jordan Journal of Biological Sciences,* 5(3).

Badgujar, S.B., 2011. *Proteolytic enzymes of some latex bearing plants belonging to Khandesh region of Maharashtra* [Ph. D. thesis]. North Maharashtra University, Jalgaon (Maharashtra State), India.

Badgujar, S.B., Patel, V.V., Bandivdekar, A.H. and Mahajan, R.T., 2014. Traditional uses, phytochemistry and pharmacology of *Ficus carica*: A review. *Pharmaceutical Biology,* 52(11), pp.1487-1503.

Bae, S.-H., Suh, H.-J., 2007. Antioxidant activities of five different mulberry cultivars in Korea. *LWT-Food Science and Technology.* 40: 955–962.

Bahuguna, V.K., n.d. Hilaluddin. 1995 Plant community classification and ordination of sal forests of Bankura (north) forest division of West Bengal, India. *Van Vigyan* 33, 87–103.

Balta, M.F., Yarılgaç, T., Aşkın, M.A., Kuçuk, M., Balta, F., Özrenk, K., 2006. Determination of fatty acid compositions, oil contents and some quality traits of hazelnut genetic resources grown in eastern Anatolia of Turkey. *J. Food Compos. Anal.* 19, 681–686.

Banerjee, S.K., Mishra, T.K., 1996. Extension of forestry research through Joint Forest Management: An ecological impact study. *Adv. For. Res. India* 14, 11–25.

Batsatsashvili, K., Mehdiyeva, N., Kikvidze, Z., Khutsishvili, M., Maisaia, I., Sikharulidze, S., Tchelidze, D., Alizade, V., Zambrana, N.Y.P., Bussmann, R.W., 2017. *Matteuccia struthiopteris* (L.) Todd. *Ethnobot. Caucasus.* Cham Springer Int. Publ.

Belguith-Hadriche, O., Ammar, S., del Mar Contreras, M., Turki, M., Segura-Carretero, A., El Feki, A., Makni-Ayedi, F. and Bouaziz, M., 2016. Antihyperlipidemic and antioxidant activities of edible Tunisian *Ficus carica* L. fruits in high fat diet-induced hyperlipidemic rats. *Plant Foods for Human Nutrition,* 71(2), pp.183-189.

Berg, C.C., 2003. Flora Malesiana precursor for the treatment of Moraceae 1: the main subdivision of Ficus: the subgenera. *Blumea-Biodiversity, Evolution and Biogeography of Plants*, 48(1), pp.166-177.

Bhandary, M.R., Kawabata, J., 2008. Antidiabetic activity of Laligurans (*Rhododendron arboreum* Sm.) flower. *J. Food Sci. Technol. Nepal* 4, 61–63.

Bhat, M.Z.A., Ali, D.M. and Mir, S.R., 2013. Anti-diabetic activity of *Ficus carica* L. stem barks and isolation of two new flavonol esters from the plant by using spectroscopical techniques. *Asian J Biomed Pharm Sci*, 3(18), pp.22-28.

Bhat, S.D., Rawat, G.S., 1995. Habitat use by chital (Axis axis) in Dhaulkhand, Rajaji National Park, India. *Trop. Ecol.* 36, 177–189.

Bhatia, A., Meena, B., Shukla, S.K., Sidhu, O.P., Upreti, D.K., Mishra, A., Roy, R. and Nautiyal, C.S., 2017. Determination of Pentacyclic Triterpenes from *Betula utilis* by High-Performance Liquid Chromatography and High-Resolution Magic Angle Spinning Nuclear Magnetic Resonance Spectroscopy. *Analytical Letters*, 50, 233–242.

Bhatnagar, P., Hardaha, S., 1994. Socio-economic potential of non-wood forest products: a case study of Karanjia block. *Vaniki Sandesh* 18, 11–22.

Bhattacharyya, A., Shah, S.K., Chaudhary, V., 2006. Would tree ring data of *Betula utilis* be potential for the analysis of Himalayan glacial fluctuations? *Current Science*, 754–761.

Bhattacharyya, D., 2011. Rhododendron species and their uses with special reference to Himalayas–a review. *Assam Univ. J. Sci. Technol.* 7, 161–167.

Biswas, K., 1956. *Common Medicinal Plants of Darjeeling and the Sikkim Himalayas.* Alipore, West Bengal, India : Supt., Govt. Print., West Bengal Govt. Press.

Bobrowski, M., Gerlitz, L., and Schickhoff, U. 2017. Modelling the potential distribution of *Betula utilis* in the Himalaya. *Global Ecology and Conservation*,11, 69–83.

Boccacci, P., Aramini, M., Valentini, N., Bacchetta, L., Rovira, M., Drogoudi, P., Silva, A.P., Solar, A., Calizzano, F., Erdoğan, V., 2013. Molecular and morphological diversity of on-farm hazelnut (*Corylus avellana* L.) landraces from southern Europe and their role in the origin and diffusion of cultivated germplasm. *Tree Genet. Genomes* 9, 1465–1480.

Boshtam, M., Rafiei, M., Sadeghi, K., Sarraf-Zadegan, N., 2002. Vitamin E can reduce blood pressure in mild hypertensives. *Int. J. Vitam. Nutr. Res.* 72, 309–314.

Budantseva, A.L., 1996. *Plant Resources of Russia and Neighboring Countries,* vols. 1–2. Moscow Russ. Acad. Sci.

Burkill, I.H., 1935. *A Dictionary of the Economic Products of the Malay Peninsula.* Volume II (IZ). London: Crown Agents for the Colonies.

Bussmann, R.W., Paniagua, Z., Narel, Y., Sikharulidze, S., Kikvidze, Z., Kikodze, D., Tchelidze, D., Batsatsashvili, K., Robbie, E., 2017a. *Plants in the Spa–The Medicinal Plant Market of Borjomi,* Sakartvelo (Republic of Georgia), Caucasus.

Bussmann, R.W., Zambrana, N.Y.P., Sikharulidze, S., Kikvidze, Z., Kikodze, D., Jinjikhadze, T., Shanshiashvili, T., Chelidze, D., Batsatsashvili, K., Bakanidze, N., 2014. Wine, Beer, Snuff, Medicine, and Loss of Diversity-Ethnobotanical Travels in the Georgian Caucasus. *Ethnobotany Research and Applications* 12.

Bussmann, R.W., Zambrana, N.Y.P., Sikharulidze, S., Kikvidze, Z., Kikodze, D., Tchelidze, D., Khutsishvili, M., Batsatsashvili, K., Hart, R.E., 2016. A comparative ethnobotany of Khevsureti, Samtskhe-Javakheti, Tusheti, Svaneti, and Racha-Lechkhumi, Republic of Georgia (Sakartvelo), Caucasus. *J. Ethnobiol. Ethnomed.* 12, 43.

Bussmann, R.W., Zambrana, P., Narel, Y., Sikharulidze, S., Kikvidze, Z., Kikodze, D., Tchelidze, D., Batsatsashvili, K., Robbie, E., 2017b. *Ethnobotany of Samtskhe-Javakheti*, Sakartvelo (Republic of Georgia), Caucasus.

Butt, M.S., Nazir, A., Sultan, M.T., Schroën, K., 2008. *Morus alba* L. nature's functional tonic. *Trends in Food Science & Technology.* 19, 505–512.

Çalişkan, O. and Polat, A.A., 2011. Phytochemical and antioxidant properties of selected fig (*Ficus carica* L.) accessions from the eastern Mediterranean region of Turkey. *Scientia Horticulturae,* 128(4), pp.473-478.

Camero, M., Marinaro, M., Lovero, A., Elia, G., Losurdo, M., Buonavoglia, C. and Tempesta, M., 2014. In vitro antiviral activity of *Ficus carica* latex against caprine herpesvirus-1. *Natural Product Research,* 28(22), pp.2031-2035.

Canal, J.R., Torres, M.D., Romero, A. and Pérez, C., 2000. A chloroform extract obtained from a decoction of *Ficus carica* leaves improves the cholesterolaemic status of rats with streptozotocin-induced diabetes. *Acta Physiologica Hungarica,* 87(1), pp.71-76.

Carlton, G.C., and Bazzaz, F.A. 1998. Regeneration of three sympatric birch species on experimental hurricane blowdown microsites. *Ecological Monographs,* 68, 99–120.

Catoni, R., Granata, M.U., Sartori, F., Varone, L., Gratani, L., 2015. *Corylus avellana* responsiveness to light variations: morphological, anatomical, and physiological leaf trait plasticity. *Photosynthetica* 53, 35–46.

Celep, E., Charehsaz, M., Akyüz, S., Acar, E.T., Yesilada, E., 2015. Effect of in vitro gastrointestinal digestion on the bioavailability of phenolic components and the antioxidant potentials of some Turkish fruit wines. *Food Research International,* 78, 209–215.

Cerulli, A., Lauro, G., Masullo, M., Cantone, V., Olas, B., Kontek, B., Nazzaro, F., Bifulco, G., Piacente, S., 2017. Cyclic diarylheptanoids from *Corylus avellana* green leafy covers: determination of their absolute configurations and evaluation of their antioxidant and antimicrobial activities. *J. Nat. Prod.* 80, 1703–1713.

Cerulli, A., Masullo, M., Montoro, P., Hošek, J., Pizza, C., Piacente, S., 2018a. Metabolite profiling of "green" extracts of *Corylus avellana* leaves by 1H NMR spectroscopy and multivariate statistical analysis. *J. Pharm. Biomed. Anal.* 160, 168–178.

Cerulli, A., Napolitano, A., Masullo, M., Pizza, C., Piacente, S., 2018b. LC-ESI/LTQOrbitrap/MS/MSn analysis reveals diarylheptanoids and flavonol O-glycosides in fresh and roasted hazelnut (*Corylus avellana* cultivar "Tonda di Giffoni"). *Nat. Prod. Commun.* 13, 1934578X1801300906.

Champion, H.G., Seth, S.K., 1968. *General Silviculture for India.* Natraj.

Chan, E.W., Lye, P.-Y., Wong, S.-K., 2016. Phytochemistry, pharmacology, and clinical trials of *Morus alba. Chinese Journal of Natural Medicines,* 17–30.

Chandra, P.K. 2010. *Medicinal Plants of Uttarakhand: Diversity, Livelihood and Conservation.* Biotech Books.

References

Chauhan, P., Singh, J., Sharma, R.K., Easwari, T.S., 2016. Anti-bacterial activity of *Rhododendron arboreum* plant against Staphylococcus aureus. *Ann. Hortic.* 9, 92–96.

Chauhan, S.M.S., Singh, M., Narayan, L., 2002. Isolation of 3β-hydroxyolean-12-ene, friedelin and 7-methoxy-4'-5-dihydroxyisoflavone from Dry and Fresh Leaves of Shorea robusta. *Indian Journal of Chemistry Section B*, 41(5):1097-1099.

Chawla, A., Kaur, R. and Sharma, A.K., 2012. Ficus carica Linn.: A review on its pharmacognostic, phytochemical and pharmacological aspects. *International Journal of Pharmaceutical and Phytopharmacological Research*, 1(4), pp.215-232.

Chen, C.-C., Liu, L.-K., Hsu, J.-D., Huang, H.-P., Yang, M.-Y., Wang, C.-J., 2005. Mulberry extract inhibits the development of atherosclerosis in cholesterol-fed rabbits. *Food Chemistry.* 91, 601–607.

Chetri, R., Pandey, T., 1992. *User Group Forestry in the Far-Western Region of Nepal.* Kathmandu, ICIMOD.

Chinese Pharmacopeia Commission. *Pharmacopoeia of the People's Republic of China.* Chinese Medical Science and Technology Press: Beijing, China, 2015; Vol. 1, p 300

Cho, E., Y Chung, E., Jang, H.-Y., Hong, O.-Y., S Chae, H., Jeong, Y.-J., Kim, S.-Y., Kim, B.-S., J Yoo, D., Kim, J.-S., 2017. Anti-cancer effect of cyanidin-3-glucoside from mulberry via caspase-3 cleavage and DNA fragmentation in vitro and in vivo. *Anti-Cancer Agents in Medicinal Chemistry (Formerly Current Medicinal Chemistry-Anti-Cancer Agents)*, 17, 1519–1525.

Chon, S.-U., Kim, Y.-M., Park, Y.-J., Heo, B.-G., Park, Y.-S., Gorinstein, S., 2009. Antioxidant and antiproliferative effects of methanol extracts from raw and fermented parts of mulberry plant (*Morus alba* L.). *European Food Research and Technology.* 230, 231–237.

Chopra, S., Chatterjee, D., 1990. Integrating conservation and development: a case study of the socio-economic forestry complex at Arabari, West Bengal. *Int. Tree Crop. J.* 6, 193–204.

Ciarmiello, L.F., Mazzeo, M.F., Minasi, P., Peluso, A., De Luca, A., Piccirillo, P., Siciliano, R.A., Carbone, V., 2014. Analysis of different European hazelnut (*Corylus avellana* L.) cultivars: authentication, phenotypic features, and phenolic profiles. *J. Agric. Food Chem.* 62, 6236–6246.

Cichewicz, R.H., and Kouzi, S.A. 2004. Chemistry, biological activity, and chemotherapeutic potential of betulinic acid for the prevention and treatment of cancer and HIV infection. *Medicinal Research Reviews*, 24, 90–114.

Clinovschi, F., 2005. *Dendrologie [Dendrology].* Editura Universității din Suceava.

Cristofori, V., Ferramondo, S., Bertazza, G., Bignami, C., 2008. Nut and kernel traits and chemical composition of hazelnut (*Corylus avellana* L.) cultivars. *J. Sci. Food Agric.* 88, 1091–1098.

Crosby, A.W., and Worster, D. 1998. *Sediments of Time: Environment and Society in Chinese History.* Cambridge University Press.

Cummings, J.H., Macfarlane, G.T., 2002. Gastrointestinal effects of prebiotics. *British Journal of Nutrition*, 87, S145–S151.

Dabili, S., Fallah, S., Aein, M., Vatannejad, A., Panahi, G., Fadaei, R., Moradi, N., Shojaii, A., 2019. Survey of the effect of doxorubicin and flavonoid extract of white

Morus alba leaf on apoptosis induction in a-172 GBM cell line. *Archives of Physiology and Biochemistry.* 125, 136–141.

Damirov, I.A., Prilipko, L.I., Shukurov, D.Z., Kerimov, Y.B., 1988. *Medicinal Plants of Azerbaijan* (in Russian).

Dangwal, L.R., Singh, T., Singh, A., 2014. Exploration of wild edible plants used by Gujjar and Bakerwal tribes of District Rajouri (J&K), India. *J. Appl. Nat. Sci.* 6, 164–169.

Dat, N.T., Binh, P.T.X., Van Minh, C., Huong, H.T., Lee, J.J., 2010. Cytotoxic prenylated flavonoids from *Morus alba*. *Fitoterapia* 81, 1224–1227.

de Amorin, A., Borba, H.R., Carauta, J.P., Lopes, D. and Kaplan, M.A., 1999. Anthelmintic activity of the latex of Ficus species. *Journal of Ethnopharmacology,* 64(3), pp.255-258.

De Jong, P.C. 1993. "An introduction to *Betula*: its morphology, evolution, classification and distribution, with a survey of recent work", in: *Proceedings of the IDS Betula Symposium,* 1992. International Dendrology Society.

de Milleville, R., 2002. *The Rhododendrons of Nepal.* Himal Books.

Debib, A., Tir-Touil, M.A., Meddah, B., Hamaidi-Chergui, F., Menadi, S. and Alsayadi, M.S., 2018. Evaluation of antimicrobial and antioxidant activities of oily macerates of Algerian dried figs (*Ficus carica* L.). *International Food Research Journal,* 25(1).

Debprasad, C., Hemanta, M., Paromita, B., Durbadal, O., Kumar, K.A., Shanta, D., Kumar, H.P., Tapan, C., Ashoke, S., Sekhar, C., 2012. Inhibition of, TNF-and iNOS EXpression by *Shorea robusta* L.: An Ethnomedicine Used for Anti-Inflammatory and Analgesic Activity. Evidence-Based Complement. *Altern. Med.* 2012.

Deepa, M., Sureshkumar, T., Satheeshkumar, P.K., Priya, S., 2013. Antioxidant rich *Morus alba* leaf extract induces apoptosis in human colon and breast cancer cells by the downregulation of nitric oxide produced by inducible nitric oxide synthase. *Nutrition and Cancer* 65, 305–310.

Del Rio, D., Calani, L., Dall'Asta, M., Brighenti, F., 2011. Polyphenolic composition of hazelnut skin. *J. Agric. Food Chem.* 59, 9935–9941.

Delgado, T., Malheiro, R., Pereira, J.A., Ramalhosa, E., 2010. Hazelnut (*Corylus avellana* L.) kernels as a source of antioxidants and their potential in relation to other nuts. *Ind. Crops Prod.* 32, 621–626.

Demigné, C., Sabboh, H., Rémésy, C., Meneton, P., 2004. Protective effects of high dietary potassium: nutritional and metabolic aspects. *J. Nutr.* 134, 2903–2906.

Demirbas, A., 2007. Phenolics from hazelnut kernels by supercritical methanol extraction. *Energy Sources,* Part A 29, 791–797.

Devi, S., Vats, C.K., Dhaliwal, Y.S., 2018. Quality evaluation of *Rhododendron arboreum* flowers of different regions of Himachal Pradesh for standardization of juice extraction technique. *Int. J. Adv. Agric. Sci. Technol.* 5, 51–57.

Dey, A., Gupta, B., De, J.N., 2012. Traditional phytotherapy against skin diseases and in wound healing of the tribes of Purulia district, West Bengal, India. *J. Med. Plants Res.* 6, 4483–4825.

Dhein, S., Kabat, A., Olbrich, A., Rösen, P., Schröder, H., Mohr, F.-W., 2003. Effect of chronic treatment with vitamin E on endothelial dysfunction in a type I in vivo diabetes mellitus model and in vitro. *J. Pharmacol. Exp. Ther.* 305, 114–122.

Dinerstein, E., 1979. An ecological survey of the Royal Karnali-Bardia Wildlife Reserve, Nepal. Part I: vegetation, modifying factors, and successional relationships. *Biol. Conserv.* 15, 127–150.

Doi, K., Kojima, T., Fujimoto, Y., 2000. Mulberry leaf extract inhibits the oxidative modification of rabbit and human low density lipoprotein. *Biological and Pharmaceutical Bulletin*. 23, 1066–1071.

Doi, K., Kojima, T., Makino, M., Kimura, Y., Fujimoto, Y., 2001. Studies on the constituents of the leaves of *Morus alba* L. *Chemical and Pharmaceutical Bulletin*. 49, 151–153.

Doro, B., Garsa, S.M.B., Abusua, F.I.M., Elmarbet, N. and Shaban, M.B., 2018. In vitro Antibacterial Activity of *Ficus carica* L. (Moraceae) from Libya. *Journal of Complementary and Alternative Medical Research*, pp.1-8.

Du, J., He, Z.-D., Jiang, R.-W., Ye, W.-C., Xu, H.-X., But, P.P.-H., 2003. Antiviral flavonoids from the root bark of *Morus alba* L. *Phytochemistry* 62, 1235–1238.

Du, Q.-Z., Zheng, J., Xu, Y., 2008. Composition of anthocyanins in mulberry and their antioxidant activity. *Journal of Food Composition and Analysis*. 21, 390–395.

Duddukuri, G.R., Rao, D.E., Kaladhar, D., Sastry, N., Rao, K.K., Chaitanya, K.K., Sireesha, C., 2011. Preliminary studies on in vitro antibacterial activity and phytochemical analysis of aqueous crude extract of *Shorea robusta* floral parts. *Int J Curr Res* 3, 21–23.

Dueñas, M., Pérez-Alonso, J.J., Santos-Buelga, C. and Escribano-Bailón, T., 2008. Anthocyanin composition in fig (*Ficus carica* L.). *Journal of Food Composition and Analysis*, 21(2), pp.107-115.

Duke JA, Bugenschutz-godwin MJ, J Du collier, and P. K. Duke, *Hand Book of Medicinal Herbs*. CRC Press, Boca Raton, Fla, USA, 2nd edition, 2002.

Durak, İ., 1999. Hazelnut supplementation enhances plasma antioxidant potential and lowers plasma cholesterol levels. *Clin. Chem. Acta* 284, 113–115.

Dwivedi, A.P., 1993. Forests: the non-wood resources. *International Book Distributors*.

Dzubak, P., Hajduch, M., Vydra, D., Hustova, A., Kvasnica, M., Biedermann, D., Markova, L., Urban, M., and Sarek, J. 2006. Pharmacological activities of natural triterpenoids and their therapeutic implications. *Natural Product Reports*, 23, 394–411.

Edwards, D.M., 1996. Non-timber forest products from Nepal. *FORESC Monogr. 1/96*.

El-Beshbishy, H.A., Singab, A.N.B., Sinkkonen, J., Pihlaja, K., 2006. Hypolipidemic and antioxidant effects of *Morus alba* L.(Egyptian mulberry) root bark fractions supplementation in cholesterol-fed rats. *Life Sciences*. 78, 2724–2733.

El-Shobaki, F.A., El-Bahay, A.M., Esmail, R.S.A., El-Megeid, A.A.A. and Esmail, N.S., 2010. Effect of figs fruit (*Ficus carica* L.) and its leaves on hyperglycemia in alloxan diabetic rats. *World Journal of Dairy & Food Sciences*, 5(1), pp.47-57.

Emniyet, A.A., Avci, E., Ozcelik, B., Avci, G.A., Kose, D.A., 2014. Antioxidant and antimicrobial activities with GC/MS analysis of the *Morus alba* L. leaves. *Hittite Journal of Science and Engineering*. 1: 37–41

Enkhmaa, B., Shiwaku, K., Katsube, T., Kitajima, K., Anuurad, E., Yamasaki, M., Yamane, Y., 2005. Mulberry (*Morus alba* L.) leaves and their major flavonol quercetin 3-(6-malonylglucoside) attenuate atherosclerotic lesion development in LDL receptor-deficient mice. *The Journal of Nutrition* 135, 729–734.

Eo, H.J., Park, J.H., Park, G.H., Lee, M.H., Lee, J.R., Koo, J.S., Jeong, J.B., 2014. Anti-inflammatory and anti-cancer activity of mulberry (*Morus alba* L.) root bark. BMC Complement. *BMC Complementary and Alternative Medicine*. 14, 200.

Ercisli, S., Orhan, E., 2007. Chemical composition of white (*Morus alba*), red (*Morus rubra*) and black (*Morus nigra*) mulberry fruits. *Food Chemistry*. 103, 1380–1384.

Esposito, T., Sansone, F., Franceschelli, S., Del Gaudio, P., Picerno, P., Aquino, R.P., Mencherini, T., 2017. Hazelnut (*Corylus avellana* L.) shells extract: phenolic composition, antioxidant effect and cytotoxic activity on human cancer cell lines. *Int. J. Mol. Sci.* 18, 392.

Eteraf-Oskouei, T., Allahyari, S., Akbarzadeh-Atashkhosrow, A., Delazar, A., Pashaii, M., Gan, S.H. and Najafi, M., 2015. Methanolic extract of *Ficus carica* Linn. leaves exerts antiangiogenesis effects based on the rat air pouch model of inflammation. *Evidence-Based Complementary and Alternative Medicine*, 2015.

Fallah, S., Hajihassan, Z., Zarkar, N., Chadegani, A.R., Mohammadnejad, J., Hajimirzamohammad, M., 2018. Evaluation of anticancer activity of extracted flavonoids from *Morus alba* leaves and its interaction with DNA. *Brazilian Archives of Biology and Technology*. 61.

Fallico, B., Arena, E., Zappala, M., 2003. Roasting of hazelnuts. Role of oil in colour development and hydroxymethylfurfural formation. *Food Chem.* 81, 569–573.

Fanali, C., Tripodo, G., Russo, M., Della Posta, S., Pasqualetti, V., De Gara, L., 2018. Effect of solvent on the extraction of phenolic compounds and antioxidant capacity of hazelnut kernel. *Electrophoresis* 39, 1683–1691.

Fathy, S.A., Singab, A.N.B., Agwa, S.A., El Hamid, D.M.A., Zahra, F.A., El Moneim, S.M.A., 2013. The antiproliferative effect of mulberry (*Morus alba* L.) plant on hepatocarcinoma cell line HepG2. *Egyptian Journal of Medical Human Genetics*. 14, 375–382.

Fox, J., 1995. *Society and Non-Timber Forest Products in Tropical Asia*. Honolulu: East-West Center.

Francis, S.V., Senapati, S.K., Rout, G.R., 2007. Rapid clonal propagation of Curculigo orchioides Gaertn., an endangered medicinal plant. *Vitr. Cell. Dev. Biol.* 43, 140–143.

Fu, Q.H., 1994. Dipterocarp trees in Hinan Island, South China. *Trop. For.* 55–59.

Gabrielyan, E., Amroyan, E., Zakaryan, A., Grigoryan, G., 2004. The regulation of herbal medicinal products in Armenia. *Drug Inf. J.* 38, 273–281.

Gadgil, M., 1990. India's deforestation: patterns and processes. *Soc. Nat. Resour.* 3, 131–143.

Gadgil, M., Guha, R., 1993. *This Fissured Land: An Ecological History of India*. Univ of California Press.

Gallego Palacios, A., 2015. *Corylus avellana: A New Biotechnological Source of Anticancer Agents*.

Gammerman, A.F., Grom, I.I., 1976. *Wild Medicinal Plants of the USSR*. Moscow Russ. Acad. Sci.

Ganeshaiah, K.N., Uma Shaanker, R., Murali, K.S., n.d. Uma Shaankar, and Bawa KS., 1998,"Extraction of non-timber forest products in the forests of Biligiri Rangan Hills, India. 5 influence of dispersal mode on species response to anthropogenic pressures." *Econ. Bot* 52, 316–319.

Gangopadhyay, S.K., Nath, S., Das, P.K., Banerjee, S.K., 1990. Distribution of organic matter in coppice sal (*Shorea robusta*) in relation to soil chemical attributes. *Indian For.* 116, 407–417.

Garcia, J.M., Agar, I.T., Streif, J., 1994. Lipid characteristics of kernels from different hazelnut varieties. *Turkish J. Agric. For.* 18, 199–202.

Gautam, K.H., 1990. *Regeneration status of sal (Shorea robusta) forests in Nepal.* Dep. For. Nepal, Kathmandu 11.

Gautam, K.H., 1991. *Indigenous Forest Management Systems in the Hills of Nepal.*

Gautam, K.H., Devkota, B.P., 1999. Sal (*Shorea robusta*) leaves can provide income to some community forestry user groups at Sindhupalchok district. *Nepal J.* 11, 39–46.

Gautam, S., Chaudhary, K., 2020. Evaluation of anti-diabetic and anti-hyperlipidemic activity of *Rhododendron arboreum* bark extract against streptozocin induced diabetes in rats. *J. Med. Herbs Ethnomedicine* 11–16.

Gautam, V., Kohli, S.K., Arora, S., Bhardwaj, R., Kazi, M., Ahmad, A., Raish, M., Ganaie, M.A., Ahmad, P., 2018. Antioxidant and antimutagenic activities of different fractions from the leaves of *Rhododendron arboreum* Sm. and their GC-MS profiling. *Molecules* 23, 2239.

Gautam, V., Sharma, A., Arora, S., Bhardwaj, R., 2016. Bioactive compounds in the different extracts of flowers of *Rhododendron arboreum* Sm. *J. Chem. Pharm. Res.* 8, 439–444.

Gautam, V., Sharma, A., Arora, S., Bhardwaj, R., Ahmad, A., Ahamad, B., Ahmad, P., 2020. In-vitro antioxidant, antimutagenic and cancer cell growth inhibition activities of *Rhododendron arboreum* leaves and flowers. *Saudi J. Biol. Sci.*

Genç, G.E. and Özhatay, N., 2006. An ethnobotanical study in Çatalca (European part of Istanbul) II. *Turkish Journal of Pharmaceutical Sciences,* 3(2), pp.73-89.

Genovese, C., Scarpaci, K.S., D'angeli, F., Caserta, G., Nicolosi, D., 2020. Annals of Clinical Case Studies. *Stud* 2, 1017.

Ghanbari, A., Le Gresley, A., Naughton, D., Kuhnert, N., Sirbu, D. and Ashrafi, G.H., 2019. Biological activities of *Ficus carica* latex for potential therapeutics in human papillomavirus (HPV) related cervical cancers. *Scientific Reports,* 9(1), pp.1-11.

Ghazanfar, S.A. and Al-Al-Sabahi, A.M., 1993. Medicinal plants of northern and central Oman (Arabia). *Economic Botany,* 47(1), pp.89-98.

Giampieri, F., Tulipani, S., Alvarez-Suarez, J.M., Quiles, J.L., Mezzetti, B., Battino, M., 2012. The strawberry: composition, nutritional quality, and impact on human health. *Nutrition* 28, 9–19.

Gibson, G.R., Fuller, R., 2000. Aspects of in vitro and in vivo research approaches directed toward identifying probiotics and prebiotics for human use. *The Journal of Nutrition.* 130, 391S-395S.

Gilani, A.H., Mehmood, M.H., Janbaz, K.H., Khan, A.U. and Saeed, S.A., 2008. Ethnopharmacological studies on antispasmodic and antiplatelet activities of *Ficus carica*. *Journal of Ethnopharmacology,* 119(1), pp.1-5.

Gill, S., Panthari, P., Kharkwal, H., 2015. Phytochemical investigation of high altitude medicinal plants *Cinnamomum tamala* (Buch-ham) Nees and Eberm and *Rhododendron arboreum* Smith. *Am J Phytomed Clin Ther* 3, 512–528.

Gilmour, D.A., Fisher, R.J., 1991. *Villagers, Forests, and Foresters: The Philosophy, Process, and Practice of Community Forestry in Nepal.*

Giriraj, A., Irfan-Ullah, M., Ramesh, B.R., Karunakaran, P. V, Jentsch, A., Murthy, M.S.R., 2008. Mapping the potential distribution of *Rhododendron arboreum* Sm. ssp. nilagiricum (Zenker) Tagg (Ericaceae), an endemic plant using ecological niche modelling. *Curr. Sci.* 1605–1612.

Gond, N.Y. and Khadabadi, S.S., 2008. Hepatoprotective activity of *Ficus carica* leaf extract on rifampicin-induced hepatic damage in rats. *Indian Journal of Pharmaceutical Sciences,* 70(3), p.364.

Gorji, N., Moeini, R., Memariani, Z., 2018. Almond, hazelnut and walnut, three nuts for neuroprotection in Alzheimer's disease: A neuropharmacological review of their bioactive constituents. *Pharmacol. Res.* 129, 115–127.

Grodstein, F., Chen, J., Willett, W.C., 2003. High-dose antioxidant supplements and cognitive function in community-dwelling elderly women. *Am. J. Clin. Nutr.* 77, 975–984.

Grossheim, A.A., 1952. *Plant Richness of the Caucasus.* Moscow Russ. Acad. Sci.

Gubanov, I., Krilova, I., Tikhonova, V., 1976. *Wild Useful Plants of the USSR.* Moscow Russ. Acad. Sci.

Gültekin-Özgüven, M., Karadağ, A., Duman, Ş., Özkal, B., Özçelik, B., 2016. Fortification of dark chocolate with spray dried black mulberry (*Morus nigra*) waste extract encapsulated in chitosan-coated liposomes and bioaccessibility studies. *Food Chemistry.* 201, 205–212.

Gundogdu, M., Muradoglu, F., Sensoy, R.I.G., Yilmaz, H., 2011. Determination of fruit chemical properties of *Morus nigra* L., *Morus alba* L. and *Morus rubra* L. by HPLC *Scientia horticulturae.* 132, 37–41.

Gunes, N.T., Köksal, A.İ., Artık, N., Poyrazoğlu, E., 2010. Biochemical content of hazelnut (*Corylus avellana* L.) cultivars from West Black Sea Region of Turkey. *Eur. J. Hortic. Sci.* 75, 77–84.

Gupta, R.K., Chauhan, A., West, N.E., 1996. Restoration of degraded rangelands and biodiversity of the Siwalik Hills between the Ganga and Yamuna rivers, India. Rangelands in a sustainable biosphere, in: *Fifth International Rangeland Congress,* Salt Lake City, Utah, USA.

Hagman, M. 1972. On self-and cross-incompatibility shown by *Betula verrucosa* Ehrh. and *Betula pubescens* Ehrh. *Finland Metsantutkimuslaitos Julkaisuja.*

Han, X., Shen, T., Lou, H., 2007. Dietary polyphenols and their biological significance. *Int. J. Mol. Sci.* 8, 950–988.

Haq, F. 2012. The critically endangered flora and fauna of district Battagram Pakistan. *Advances in Life Sciences,* 2, 118–123.

Haq, F., 2012. The ethno botanical uses of medicinal plants of Allai Valley, Western Himalaya Pakistan. *Int J Plant Res* 2, 21–34.

Harborne, J.B., 2013. *The Flavonoids: Advances in Research Since 1980.* Springer.

Hariharan, V., Rangaswami, S., 1966. Chemical investigation of the bark of *Rhododendron arboreum* Sm. *Curr. Sci.* 35, 390–391.

Harzallah, A., Bhouri, A.M., Amri, Z., Soltana, H. and Hammami, M., 2016. Phytochemical content and antioxidant activity of different fruit parts juices of three figs (*Ficus carica* L.) varieties grown in Tunisia. *Industrial Crops and Products*, 83, pp.255-267.

Hashemi, S.A. and Abediankenari, S., 2013. Suppressive effect of fig (*Ficus carica*) latex on esophageal cancer cell proliferation. *Acta Facultatis Medicae Naissensis*, 30(2), pp.93-96.

Hashemi, S.A., Abediankenari, S., Ghasemi, M., Azadbakht, M., Yousefzadeh, Y. and Dehpour, A.A., 2011. The effect of fig tree latex (*Ficus carica*) on stomach cancer line. *Iranian Red Crescent Medical Journal*, 13(4), p.272.

Hesham, A., Abdel, N.B., Jari, S., Kalevi, P., 2005. Hypolipidemic and antioxidant effect of *Morus alba* L (Egyptian mulberry) root bark fractions supplementation in cholesterol-fed rats. *Journal of Ethnopharmacology* 78, 2724–2733.

Hisayoshi, K., Masashi, Y., Koichi, S., 2004. A novel cytotoxic prenylated flavonoid from the root of *Morus alba*. *Journal of Insect Biotechnology and Sericology* 73, 113–116.

Hjelmroos, M. 1991. Evidence of long-distance transport of Betula pollen. *Grana*, 30, 215–228.

HMG/ADB/FINNIDA, 1988. *Master Plan for Forestry Sector Nepal*. Ministry of Forest and Soil Conservation, Kathmandu, Nepal (Main Report)

Hora, F.B., 1981. *Oxford Encyclopedia of Trees of the World*. Oxford University Press.

Hota, R.K., Bapuji, M., 1993. Triterpenoids from the resin of *Shorea robusta*. *Phytochemistry* 32, 466–468.

Hota, R.K., Bapuji, M., 1994. Triterpenoids from the resin of *Shorea robusta*. *Phytochemistry* 35, 1073–1074.

Hoxha, L., Kongoli, R. and Hoxha, M., 2015. Antioxidant activity of some dried autochthonous Albanian fig (*Ficus carica*) Cultivars. *IJCST*, 1(2).

Hsu, L.-S., Ho, H.-H., Lin, M.-C., Chyau, C.-C., Peng, J.-S., Wang, C.-J., 2012. Mulberry water extracts (MWEs) ameliorated carbon tetrachloride-induced liver damages in rat. *Food and Chemical Toxicology*. 50, 3086–3093.

Hummer, K., 2000. Historical notes on hazelnuts in Oregon, in: *V International Congress on Hazelnut* 556. pp. 25–28.

Husain, S.Z., Malik, R.N., Javaid, M. and Bibi, S.A.D.I.A., 2008. Ethonobotanical properties and uses of medicinal plants of Morgah biodiversity park, Rawalpindi. *Pak J Bot*, 40(5), pp.1897-1911.

Iannuzzi, A., Celentano, E., Panico, S., Galasso, R., Covetti, G., Sacchetti, L., Zarrilli, F., De Michele, M., Rubba, P., 2002. Dietary and circulating antioxidant vitamins in relation to carotid plaques in middle-aged women. *Am. J. Clin. Nutr.* 76, 582–587.

Ibrar, M., Hussain, F. and Sultan, A., 2007. Ethnobotanical studies on plant resources of Ranyal hills, District Shangla, Pakistan. *Pakistan Journal of Botany*, 39(2), p.329.

Idolo, M., Motti, R. and Mazzoleni, S., 2010. Ethnobotanical and phytomedicinal knowledge in a long-history protected area, the Abruzzo, Lazio and Molise National Park (Italian Apennines). *Journal of Ethnopharmacology*, 127(2), pp.379-395.

Ikram, R.M., Hussain, M.S., Khan, M.T., Ghafoor Ahamad, S.K., Khan, S.A. and Murtaza, G., 2014. Gist of medicinal plants of Pakistan having ethnobotanical evidences to crush renal calculi (kidney stones). *Acta Poloniae Pharmaceutica*, 71(1), pp.3-10.

Imran, M., Khan, H., Shah, M., Khan, R., Khan, F., 2010. Chemical composition and antioxidant activity of certain Morus species. *Journal of Zhejiang University Science B* 11, 973–980.

Ishtiaq, M., Hanif, W., Khan, M.A., Ashraf, M. and Butt, A.M., 2007. An ethnomedicinal survey and documentation of important medicinal folklore food phytonims of flora of Samahni valley, (Azad Kashmir) Pakistan. *Pakistan Journal of Biological Sciences: PJBS*, 10(13), pp.2241-2256.

Isotova, M.A., Sarafakova, N.A., Mkscho, B.I., Ionova, A.A., 2010. *Great Encyclopedia of Traditional Medicine*. Moscow Russ. Acad. Sci.

Ivanović, S., Avramović, N., Dojčinović, B., Trifunović, S., Novaković, M., Tešević, V., Mandić, B., 2020. Chemical Composition, Total Phenols and Flavonoids Contents and Antioxidant Activity as Nutritive Potential of Roasted Hazelnut Skins (*Corylus avellana* L.). *Foods* 9, 430.

Jackson, J.K., 1994. *Manual of Afforestation in Nepal*, vol. 1. Kathmandu For. Res. Surv. Centre.

Jackson, J.K., Stapleton, C.M.A., Jeanrenaud, J.-P., 1994. *Manual of Afforestation in Nepal*.

Jakopic, J., Petkovsek, M.M., Likozar, A., Solar, A., Stampar, F., Veberic, R., 2011. HPLC–MS identification of phenols in hazelnut (*Corylus avellana* L.) kernels. *Food Chem.* 124, 1100–1106.

Janzen, D.H., 1979. How to be a fig. *Annual Review of Ecology and Systematics*, 10(1), pp.13-51.

Jaradat, N.A., 2005. Medical plants utilized in Palestinian folk medicine for treatment of diabetes mellitus and cardiac diseases. *Al-Aqsa University Journal* (Natural Sciences Series), 9(1), pp.1-28.

Jegan, B.S., Selvaraj, D., 2014. Antimicrobial Activity of Ethanol Leaf Extract of *Rhododendron arboreum* ssp. nilagiricum. *J. Basic Appl. Biol* 8, 1–3.

Jeong, M.R., Kim, H.Y. and Cha, J.D., 2009. Antimicrobial activity of methanol extract from *Ficus carica* leaves against oral bacteria. *Journal of Bacteriology and Virology*, 39(2), pp.97-102.

Jeong, W.S. and Lachance, P.A., 2001. Phytosterols and fatty acids in fig (*Ficus carica*, var. Mission) fruit and tree components. *Journal of Food Science*, 66(2), pp.278-281.

Jeszka, M., Kobus-Cisowska, J., Flaczyk, E., 2009. Mulberry leaves as a source of biologically active compounds. *Postępy Fitoterapii.*. 10, 175–179.

Jeszka-Skowron, M., Flaczyk, E., Jeszka, J., Krejpcio, Z., Król, E., Buchowski, M.S., 2014. Mulberry leaf extract intake reduces hyperglycaemia in streptozotocin (STZ)-induced diabetic rats fed high-fat diet. *Journal of Functional Foods* 8, 9–17.

Jiang, Y., Nie, W.-J., 2015. Chemical properties in fruits of mulberry species from the Xinjiang province of China. *Food Chemistry*. 174, 460–466.

Jiao, Y., Wang, X., Jiang, X., Kong, F., Wang, S., Yan, C., 2017. Antidiabetic effects of *Morus alba* fruit polysaccharides on high-fat diet-and streptozotocin-induced type 2 diabetes in rats. *Journal of Ethnopharmacology*. 199, 119–127.

Jin, W.Y., 2002. Antioxidant compounds from twig of *Morus alba*. *Natural Product Sciences*. 8, 129–132.

Johnsson, H. 1974. Genetic characteristics of *Betula verrucosa* Ehrh. and *B. pubescens* Ehrh.=. *Anal. za šumarstvo*, v. 6/4.

Joint, F.A.O., 1985. Energy and protein requirements: report of a joint FAO/WHO/UNU Expert Consultation, in: *Technical Report Series* (WHO). World Health Organization.

Jolly, M.S., Sen, S.K., Das, M.G., 1976. Silk from the forest. *Unasylva* 28, 20–23.

Joshi, H., Saxena, G.K., Singh, V., Arya, E., and Singh, R.P. 2013. Phytochemical investigation, isolation and characterization of betulin from bark of *Betula utilis*. *Journal of Pharmacognosy and Phytochemistry*, 2, 145–151.

Joshi, S.K., Ballabh, B., Negi, P.S., Dwivedi, S.K., 2018. Diversity, distribution, use pattern and evaluation of wild edible plants of Uttarakhand, India. *Def. Life Sci. J.* 3, 126–135.

Jyothi, G., Carey, W.M., Kumar, R.B., Mohan, K.G., 2008. Antinociceptive and antiinflammatory activity of methanolic extract of leaves of *Shorea robusta*. *Pharmacologyonline* 1, 9–19.

Kala, C.P. 1998. *Ecology and Conservation of Alpine Meadows in the Valley of Flowers National Park, Garhwal Himalaya*. Wildl. Inst. India, Dehradun. p 99.

Kala, C.P. 2018. Uses, Population Status and Management of *Betula utilis*. *Science and Education*, 6, 79–83.

Kalaiselvan, A., Gokulakrishnan, K., 2012. Bark extract of *Shorea robusta* on modulation of immune response in rats. *Int J Recent Sci. Res* 3, 693–697.

Kalaiselvan, A., Gokulakrishnan, K., Anand, T., Akhilesh, U., Velavan, S., 2013. Preventive effect of *Shorea robusta* bark extract against diethylnitrosamine-induced hepatocellular carcinoma in rats. *Int Res J Med. Sci* 1, 2–9.

Kalaskar, M.G., Shah, D.R., Raja, N.M., Surana, S.J. and Gond, N.Y., 2010. Pharmacognostic and Phytochemical Investigation of *Ficus carica* Linn. *Ethnobotanical Leaflets*, 2010(5), p.6.

Kamil, M., Ilyas, M., 1995. Flavonoidic constituents of *Rhododendron arboreum* leaves. *Fitoter.* 66.

Kang, S.K., Chung, D.O. and Chung, H.J., 1995. Purification and identification of antimicrobial substances in phenolic fraction of fig leaves. *Applied Biological Chemistry*, 38(4), pp.293-296.

Kashyap, P., Anand, S., Thakur, A., 2017. Evaluation of antioxidant and antimicrobial activity of *Rhododendron arboreum* flowers extract. *Int. J. Food Ferment. Technol.* 7, 123–128.

Kashyap, P., Zaanand, S., 2016. Phytochemical and GC-MS analysis of *Rhododendron arboreum* flowers. *Int. J. Farm Sci.* 6, 145–151.

Katsube, T., Imawaka, N., Kawano, Y., Yamazaki, Y., Shiwaku, K., Yamane, Y., 2006. Antioxidant flavonol glycosides in mulberry (*Morus alba* L.) leaves isolated based on LDL antioxidant activity. *Food Chemistry* 97, 25–31.

Katsube, T., Tsurunaga, Y., Sugiyama, M., Furuno, T., Yamasaki, Y., 2009. Effect of air-drying temperature on antioxidant capacity and stability of polyphenolic compounds in mulberry (*Morus alba* L.) leaves. *Food Chemistry*. 113, 964–969.

Kaume, L., Howard, L.R., Devareddy, L., 2012. The blackberry fruit: a review on its composition and chemistry, metabolism and bioavailability, and health benefits. J. Agric. *Food Chemistry*. 60, 5716–5727.

Kaur, S., Varshney, V.K., Dayal, R., 2003. GC-MS analysis of essential oil of *Shorea robusta* bast. *J. Asian Nat. Prod. Res.* 5, 231–234.

Kayastha, B.P., 1985. *Silvics of the Trees of Nepal*.

Kemertelidze, E.P., Shalashvili, K.G., Korsantiya, B.M., Nizharadze, N.O., Chipashvili, N.S., 2007. Therapeutic effect of phenolic compounds isolated from Rhododendron ungernii leaves. *Pharm. Chem. J.* 41, 10–13.

Keskin, D., Ceyhan Guvensen, N., Zorlu, Z. and Ugur, A., 2012. Phytochemical analysis and antimicrobial activity of different extracts of fig leaves (*Ficus carica* L.) from West Anatolia against some pathogenic microorganisms. *J Pure Appl Microbiol*, 6, pp.1105-1110.

Khadabadi, S.S., Gond, N.Y., Ghiware, N.B. and Shendarkar, G.R., 2007. Hepatoprotective effect of *Ficus carica* leaf in chronic hepatitis. *Indian Drugs-Bombay*, 44(1), p.54.

Khan, I., Nisar, M., Ali, S., Qaisar, M., Gilani, S.N., Shah, M.R., Ali, G., 2013. Antifungal activity of bioactive constituents and bark extracts of *Rhododendron arboreum*. *Bangladesh J. Pharmacol*. 8, 218–222.

Khan, I., Sangwan, P.L., Dhar, J.K., and Koul, S. 2012. Simultaneous quantification of five marker compounds of *Betula utilis* stem bark using a validated high-performance thin-layer chromatography method. *Journal of Separation Science*, 35, 392–399.

Khan, K.Y., Khan, M.A., Ahmad, M., Mazari, P., Hussain, I., Ali, B., Fazal, H. and Khan, I.Z., 2011. Ethno-medicinal species of genus *Ficus* L. used to treat diabetes in Pakistan. *Journal of Applied Pharmaceutical Science*, 1(6), p.29.

Khan, M.A. 1975. Karachic acid: A new triterpenoid from *Betula utilis*. *Phytochemistry*, 14, 789–791.

Khan, M.A., Rahman, A.A., Islam, S., Khandokhar, P., Parvin, S., Islam, M.B., Hossain, M., Rashid, M., Sadik, G., Nasrin, S., 2013. A comparative study on the antioxidant activity of methanolic extracts from different parts of *Morus alba* L.(Moraceae). *BMC Research Notes* 6: 1–9

Khan, S.W. and Khatoon, S., 2007. Ethnobotanical studies on useful trees and shrubs of Haramosh and Bugrote valleys in Gilgit northern areas of Pakistan. *Pak J Bot*, 39(3), pp.699-710.

Khanom, F., Kayahara, H. and Tadasa, K., 2000. Superoxide-scavenging and prolyl endopeptidase inhibitory activities of Bangladeshi indigenous medicinal plants. *Bioscience, Biotechnology, and Biochemistry*, 64(4), pp.837-840.

Khare, C.P., 2008. *Indian Medicinal Plants: An Illustrated Dictionary*. Springer Science & Business Media.

Khodarahmi, G.A., Ghasemi, N., Hassanzadeh, F. and Safaie, M., 2011. Cytotoxic effects of different extracts and latex of *Ficus carica* L. on HeLa cell line. *Iranian Journal of Pharmaceutical Research: IJPR*, 10(2), p.273.

Kikuzaki, H., and Nakatani, N. 1993. Antioxidant effects of some ginger constituents. *Journal of Food Science*, 58, 1407–1410.

Kim, D., Kang, Y.M., Jin, W.Y., Sung, Y., Choi, G., Kim, H.K., 2014. Antioxidant activities and polyphenol content of *Morus alba* leaf extracts collected from varying regions. *Biomedical reports* 2, 675–680.

Kim, D.-S., Ji, H.D., Rhee, M.H., Sung, Y.-Y., Yang, W.-K., Kim, S.H., Kim, H.-K., 2014. Antiplatelet activity of *Morus alba* leaves extract, mediated via inhibiting granule secretion and blocking the phosphorylation of extracellular-signal-regulated kinase and akt. *Evidence-Based Complementary and Alternative Medicine*. 2014.

Kim, E.-S., Park, S.-J., Lee, E.-J., Kim, B.-K., Huh, H., Lee, B.-J., 1999. Purification and characterization of Moran 20K from *Morus alba*. *Archives of Pharmacal Research*. 22, 9–12.

Kim, S.B., Chang, B.Y., Hwang, B.Y., Kim, S.Y., Lee, M.K., 2014. Pyrrole alkaloids from the fruits of *Morus alba*. *Bioorganic & Medicinal Chemistry Letters*. 24, 5656–5659.

Kim, S.Y., Gao, J.J., Kang, H.K., 2000. Two flavonoids from the leaves of *Morus alba* induce differentiation of the human promyelocytic leukemia (HL-60) cell line. *Biological and Pharmaceutical Bulletin*. 23, 451–455.

Kim, S.Y., Gao, J.J., Lee, W.-C., Ryu, K.S., Lee, K.R., Kim, Y.C., 1999. Antioxidative flavonoids from the leaves of *Morus alba*. *Arch. Pharm. Res.* 22, 81.

Kirtikar, K.R. and Basu, B.D., 1995. *Indian Medicinal Plants International Book Distributors*. Deharadun, India, pp.1-456.

Kiruba, S., Mahesh, M., Nisha, S.R., Paul, Z.M., Jeeva, S., 2011. Phytochemical analysis of the flower extracts of *Rhododendron arboreum* Sm. ssp. nilagiricum (Zenker) Tagg. *Asian Pac. J. Trop. Biomed.* 1, S284–S286.

Kobus-Cisowska, J., Gramza-Michalowska, A., Kmiecik, D., Flaczyk, E., Korczak, J., 2013. Mulberry fruit as an antioxidant component in muesli. *Agric. Sci.* 4, 130.

Kobus-Cisowska, J., Szczepaniak, O., Szymanowska-Powałowska, D., Piechocka, J., Szulc, P., Dziedziński, M., 2020. Antioxidant potential of various solvent extract from *Morus alba* fruits and its major polyphenols composition. *Ciência Rural* 50.

Köksal, A.İ., Artik, N., Şimşek, A., Güneş, N., 2006. Nutrient composition of hazelnut (*Corylus avellana* L.) varieties cultivated in Turkey. *Food Chem.* 99, 509–515.

Kole, C., 2011. *Wild Crop Relatives: Genomic and Breeding Resources: Temperate Fruits*. Springer Science & Business Media.

Kopaliani, L., 2013. *Forest Plants of Georgia (Trees, Shrubs, Herbs)*. Kutais. Publ. Cent.

Kornsteiner, M., Wagner, K.-H., Elmadfa, I., 2006. Tocopherols and total phenolics in 10 different nut types. *Food Chem.* 98, 381–387.

Kull, O., Niinemets, Ü., 1993. Variations in leaf morphometry and nitrogen concentration in *Betula pendula* Roth., *Corylus avellana* L. and *Lonicera xylosteum* L. Tree Physiol. 12, 311–318.

Kumar, P., 2012. Assessment of impact of climate change on *Rhododendrons* in Sikkim Himalayas using Maxent modelling: limitations and challenges. *Biodiversity and Conservation*, 21(5), pp.1251-1266.

Kumar, V., Kumar, D., Ram, P., 2014. Varietal influence of mulberry on silkworm, Bombyx mori L. growth and development. Research Article. *Int. J. Adv. Res.* 2, 921–927.

Kumar, V., Suri, S., Prasad, R., Gat, Y., Sangma, C., Jakhu, H., Sharma, M., 2019. Bioactive compounds, health benefits and utilization of Rhododendron: a comprehensive review. *Agric. Food Secur.* 8, 6.

Kumaraswamy, M. V., and Satish, S. 2008. Free radical scavenging activity and lipoxygenase inhibition of Woodfordia fructicosa Kurz and *Betula utilis* Wall. *African Journal of Biotechnology*, 7, 2013–2016.

Kumaraswamy, M., Kavitha, H., and Satish, S. 2008. Antibacterial evaluation and phytochemical analysis of *Betula utilis* D. Don against some human pathogenic bacteria. *World Journal of Agricultural Sciences*, 2, 21–25.

Kumari, K., Srivastva, H., Mudgal, V.D., 2015. Medicinal importance and utilization of Rhododendron-A Review. *Res. Environ. Life Sci.* 8, 761–766.

Kunwar RM, Bussmann RW. (2006). Ficus (Fig) species in Nepal, a review of diversity and indigenous uses. *Lyonia* 11:85–97.

Kusano, G., Orihara, S., Tsukamoto, D., Shibano, M., Coskun, M., Guvenc, A., Erdurak, C.S., 2002. Five new nortropane alkaloids and six new amino acids from the fruit of *Morus alba* L INNÉ growing in Turkey. *Chem. Pharm. Bull.* 50, 185–192.

Lal, K., Ahuja, V., Rajeshwer, A.K.B., 2017. In Vitro Study of Antimicrobial Activity of *Rhododendron arboreum* Plant Extract on Selected Pathogenic Bacterial Isolates. *Life Sci. Int. Res.* J 4, 64–67.

Laloo, R.C., Kharlukhi, L., Jeeva, S., Mishra, B.P., 2006. Status of medicinal plants in the disturbed and the undisturbed sacred forests of Meghalaya, northeast India: population structure and regeneration efficacy of some important species. *Curr. Sci.* 225–232.

Lancaster, R., 1996. *Euphorbias: A Gardener's Guide*. Timber Press, Incorporated.

Lawesson, J.E., 2001. *Flora Nordica, Vol. 1. Lycopodiaceae to Polygonaceae*. The Bergius Foundation, Stockholm,

Lazreg Aref, H., Gaaliche, B., Fekih, A., Mars, M., Aouni, M., Pierre Chaumon, J. and Said, K., 2011. In vitro cytotoxic and antiviral activities of *Ficus carica* latex extracts. *Natural Product Research*, 25(3), pp.310-319.

Lee, A.-C., Hong, Y.-H., 2010. Development of functional yogurts prepared with mulberries and mulberry tree leaves. *Korean J. Food Sci. Ani. Resour* 30, 649–654.

Lee, S.H., Choi, S.Y., Kim, H., Hwang, J.S., Lee, B.G., Gao, J.J., Kim, S.Y., 2002. Mulberroside F isolated from the leaves of *Morus alba* inhibits melanin biosynthesis. *Biol. Pharm. Bull.* 25, 1045–1048.

Lee, Y.J., Choi, D.H., Kim, E.J., Kim, H.Y., Kwon, T.O., Kang, D.G., Lee, H.S., 2011. Hypotensive, hypolipidemic, and vascular protective effects of *Morus alba* L. in rats fed an atherogenic diet. *Am. J. Chin. Med.* 39, 39–52.

Lehmkuhl, J.F., 1994. A classification of subtropical riverine grassland and forest in Chitwan National Park, Nepal. *Vegetatio* 111, 29–43.

Lepcha, L., Basistha, B.C., Pradhan, S., Subba, K.B., Gurung, R., Sharma, N.P., 2014. Understanding Significant Value of *Rhododendron arboreum* Smith Scarleti of Sikkim, India. *Int J Eng Sci Innov Technol* 3, 554–559.

Li, H., Parry, J.W., 2011. Phytochemical compositions, antioxidant properties, and colon cancer antiproliferation effects of Turkish and Oregon hazelnut. *Food Nutr. Sci.* 2011.

Liang, L., Wu, X., Zhu, M., Zhao, W., Li, F., Zou, Y., Yang, L., 2012. Chemical composition, nutritional value, and antioxidant activities of eight mulberry cultivars from China. *Pharmacogn. Mag.* 8, 215.

Lin, C.-Y., Lay, H.-L., 2013. Characteristics of fruit growth, component analysis and antioxidant activity of mulberry (*Morus* spp.). *Sci. Hortic. (Amsterdam).* 162, 285–292.

Liu, F., Yang, Z., Zheng, X., Luo, S., Zhang, K. and Li, G., 2011. Nematicidal coumarin from *Ficus carica* L. *Journal of Asia-Pacific Entomology*, 14(1), pp.79-81.

Łochyńska, M., Oleszak, G., 2011. Multipurpose white mulberry (*Morus alba* L.).[In:] Zaikov GE, Pudel DP, Spychalski G (eds.). *Renewable Resources and Biotechnology for Material Applications.* New York : Nova Science Publishers.

Lokegaonkar, S.P., Nabar, B.M., 2011. In vitro assessment of the antiplaque properties of crude *M. alba* leaf extract. *J. Herb. Med. Toxicol* 5, 71–77.

Machii, H., Koyama, A., Yamanouchi, H., 2000. Mulberry breeding, cultivation and utilization in Japan. *FAO Anim. Prod. Heal. Pap.* 63–72.

Mahmood, T., Anwar, F., Abbas, M., Boyce, M.C., Saari, N., 2012. Compositional variation in sugars and organic acids at different maturity stages in selected small fruits from Pakistan. *Int. J. Mol. Sci.* 13, 1380–1392.

Mahmoudi, S., Khali, M., Benkhaled, A., Benamirouche, K. and Baiti, I., 2016. Phenolic and flavonoid contents, antioxidant and antimicrobial activities of leaf extracts from ten Algerian *Ficus carica* L. varieties. *Asian Pacific Journal of Tropical Biomedicine*, 6(3), pp.239-245.

Maithani, G.P., Sharma, D.C., Bahuguna, V.K., 1989. Problems of sal forests-an analysis. *Indian For.* 115, 513–525.

Manoj, A., Akhilesh, B., Ritu, K., 2014. Ayurvedic medicinal plant *Shorea robusta* (Shala). *Innovare J. Ayurvedic Sci.* 2, 18–21.

Marandi, R.R., Britto, S.J., Soreng, P.K., 2016. Phytochemical profiling, antibacterial screening and antioxidant properties of the sacred tree (*Shorea robusta* gaertn.) of Jharkhand. *Int. J. Pharm. Sci. Res.* 7, 2874.

Martin, A., 2003. Antioxidant vitamins E and C and risk of Alzheimer's disease. *Nutr. Rev.* 61, 69.

Marwat SK, Khan MA, Khan MA, et al. (2009). Fruit plant species mentioned in the Holy Qura'n and Ahadith and their ethnomedicinal importance. *Am Eurasian J Agric Environ Sci* 5:284–95.

Masullo, M., Cerulli, A., Mari, A., de Souza Santos, C.C., Pizza, C., Piacente, S., 2017. LC-MS profiling highlights hazelnut (Nocciola di Giffoni PGI) shells as a byproduct rich in antioxidant phenolics. *Food Res. Int.* 101, 180–187.

Masullo, M., Cerulli, A., Olas, B., Pizza, C., Piacente, S., 2015. Giffonins A–I, antioxidant cyclized diarylheptanoids from the leaves of the hazelnut tree (*Corylus avellana*), source of the Italian PGI Product "Nocciola di Giffoni." *J. Nat. Prod.* 78, 17–25.

Masullo, M., Mari, A., Cerulli, A., Bottone, A., Kontek, B., Olas, B., Pizza, C., Piacente, S., 2016. Quali-quantitative analysis of the phenolic fraction of the flowers of *Corylus avellana*, source of the Italian PGI product "Nocciola di Giffoni": Isolation of antioxidant diarylheptanoids. *Phytochemistry* 130, 273–281.

Mathauda, G.S., 1958. The Unevenaged Sal Forests of Ramnagar Forest Division, Uttar Pradesh: their Constitution, Rate of Growth and Drain along with Empirical Yield and Stand Tables for Selection Type of Sal Crops. *Indian For.* 84, 256–269.

Mathavi, P., Nethaji, S., 2014. In-vitro antioxidant activity of *Shorea robusta* leaf extract. *Int. J. Biochem. Biophys* 4, 22–26.

Mathema, P., 1991. Sal regeneration management. *Nepal J.* 6, 112–114.

Mawa, S., Husain, K. and Jantan, I., 2013. *Ficus carica* L.(Moraceae): phytochemistry, traditional uses and biological activities. *Evidence-Based Complementary and Alternative Medicine,* 2013.

Melkania, N.P., Ramnarayan, K., 1998. Influence of forest floor sweeping on chemical properties of lateritic soils in sal coppice forest, West Bengal, India. *Red Lateritic Soils* 1, 403–406.

Memon, A.A., Memon, N., Luthria, D.L., Bhanger, M.I., Pitafi, A.A., 2010. Phenolic acids profiling and antioxidant potential of mulberry (*Morus laevigata* W., *Morus nigra* L., *Morus alba* L.) leaves and fruits grown in Pakistan. *Polish J. Food Nutr. Sci.* 60.

Meselhy, K.M., Hammad, L.N., Farag, N., 2012. Novel antisickling, antioxidant and cytotoxic prenylated flavonoids from the bark of *Morus alba* L. *Life Sci. J.* 9.

Middelkoop, T.B., Labadie, R.P., 1983. Evaluation of Asoka Aristha, an indigenous medicine in Sri Lanka. *J. Ethnopharmacol.* 8, 313.

Middelkoop, T.B., Labadie, R.P., 1986. An in vitro Examination of Bark Extracts from *Saraca asoca* (Roxb.) de Wilde and *Rhododendron arboreum* Sm. for Oxytocic Activity. *Int. J. Crude Drug Res.* 24, 41–44.

Mishra, T., Arya, R.K., Meena, S., Joshi, P., Pal, M., Meena, B., Upreti, D.K., Rana, T.S., and Datta, D. 2016. Isolation, characterization and anticancer potential of cytotoxic triterpenes from *Betula utilis* bark. *PLoS One,* 11, e0159430.

Misra, L.N., Ahmad, A., 1997. Triterpenoids from *Shorea robusta* resin. *Phytochemistry* 45, 575–578.

Misra, R., 1969. Studies on the primary productivity of terrestrial communities at Varanasi. *Trop. Ecol.*

Mohamad, S., Zin, N.M., Wahab, H.A., Ibrahim, P., Sulaiman, S.F., Zahariluddin, A.S.M. and Noor, S.S.M., 2011. Antituberculosis potential of some ethnobotanically selected Malaysian plants. *Journal of Ethnopharmacology,* 133(3), pp.1021-1026.

Montella, R., Coïsson, J.D., Travaglia, F., Locatelli, M., Malfa, P., Martelli, A., Arlorio, M., 2013. Bioactive compounds from hazelnut skin (*Corylus avellana* L.): Effects on *Lactobacillus plantarum* P17630 and *Lactobacillus crispatus* P17631. *J. Funct. Foods* 5, 306–315.

Mopuri, R. and Islam, M.S., 2016. Antidiabetic and anti-obesity activity of *Ficus carica*: In vitro experimental studies. *Diabetes & Metabolism,* 42(4), p.300.

Mopuri, R. and Islam, M.S., 2017. Medicinal plants and phytochemicals with anti-obesogenic potentials: A review. *Biomedicine & Pharmacotherapy*, 89, pp.1442-1452.

Morrison, J., 2006. Grasslands of the World. Edited by JM Suttie, SG Reynolds and C. Batello. Rome: Food and Agriculture Organization of the United Nations, (2005), pp. 514). ISBN 92-5-105337-5. *Exp. Agric.* 42, 254.

MR Patlolla, J., and V Rao, C. 2012. Triterpenoids for cancer prevention and treatment: current status and future prospects. *Current Pharmaceutical Biotechnology*, 13, 147–155.

Mudagal, M.P., Goli, D., 2011. Preventive effect of *Rhododendron arboreum* on cardiac markers, lipid peroxides and antioxidants in normal and isoproterenol-induced myocardial necrosis in rats. *African J. Pharm. Pharmacol.* 5, 755–763.

Mujeeb, M., Khan, S.A., Aeri, V. and Ali, B., 2011. Hepatoprotective Activity of the Ethanolic Extract of *Ficus carica* Linn. LeavesinCarbon Tetrachloride-Induced Hepatotoxicityin Rats. *Iranian Journal of Pharmaceutical Research: IJPR*, 10(2), p.301.

Mukherjee, H., Ojha, D., Bharitkar, Y.P., Ghosh, S., Mondal, S., Kaity, S., Dutta, S., Samanta, A., Chatterjee, T.K., Chakrabarti, S., 2013. Evaluation of the wound healing activity of *Shorea robusta*, an Indian ethnomedicine, and its isolated constituent (s) in topical formulation. *J. Ethnopharmacol.* 149, 335–343.

Mukhopadhyay, S., 1991. Tribal area development through forest resource management: a study on southern Banduan police station of Purulia district, West Bengal. *Indian J. Reg. Sci* 23, 59–67.

Murthy, K.S.R., 2011. Biological activity and phytochemical screening of the oleoresin of *Shorea robusta* Gaertn. f. *Trop. Subtrop. Agroecosystems* 14.

Murty, D., Rajesh, E., Raghava, D., Raghavan, T.V., Surulivel, M.K.M., 2010. Hypolipidemic effect of arborium plus in experimentally induced hypercholestermic rabbits. *Yakugaku Zasshi* 130, 841–846.

Nadakarni, K. 1976. *Betula utilis* D. Don. *Indian Mater Med*, 1, 198–1296.

Nadkarni, K.M. and Nadkarni, A.K., 1976. *Indian Materia Medica* (Vol. II). Popular Prakashan Private Limited, Bombay, India.

Nadkarni, K.M., 1996. [*Indian Materia Medica*]; *Dr. KM Nadkarni's Indian materia medica: with Ayurvedic, Unani-Tibbi, Siddha, allopathic, homeopathic, naturopathic & home remedies, appendices & indexes*. 1 (Vol. 1). Popular Prakashan.

Nair, K.N.R., 1945. The problem of sal regeneration in Mayurbhanj. *Indian For.* 71, 128–129.

Nand, K., Naithani, S., 2018. Ethnobotanical uses of wild medicinal plants by the local community in the Asi Ganga Sub-basin, Western Himalaya. *J. Complement. Med. Res.* 9, 34–46.

Naowaratwattana, W., De-Eknamkul, W., De Mejia, E.G., 2010. Phenolic-containing organic extracts of mulberry (*Morus alba* L.) leaves inhibit HepG2 hepatoma cells through G2/M phase arrest, induction of apoptosis, and inhibition of topoisomerase IIα activity. *J. Med. Food* 13, 1045–1056.

Natić, M.M., Dabić, D.Č., Papetti, A., Akšić, M.M.F., Ognjanov, V., Ljubojević, M., Tešić, Ž.L., 2015. Analysis and characterisation of phytochemicals in mulberry

(*Morus alba* L.) fruits grown in Vojvodina, North Serbia. *Food Chem.* 171, 128–136.

Nayar, M.P., Ramamurthy, K., Agarwal, V.S., 1989. Economic plants of India. *Botanical Survey of India.*

Nebedum, J.O., Udeafor, P.C. and Okeke, C.U., 2010. Comparative effects of ethanolic extracts of *Ficus carica* and *Mucuna pruriens* leaves on haematological parameters in albino rats. *Biokemistri*, 22(2).

Negi, V.S., Maikhuri, R.K., Rawat, L.S., Chandra, A., 2013. Bioprospecting of *Rhododendron arboreum* for livelihood enhancement in central Himalaya, India. *Environ. We Int. Jouranl Sci. Technol.* 8, 61–70.

Nepal, S.K., Weber, K.E., 1995. Prospects for coexistence: wildlife and local people. *Ambio* 238–245.

Nisar, M., Ali, S., Muhammad, N., Gillani, S.N., Shah, M.R., Khan, H., Maione, F., 2016. Antinociceptive and anti-inflammatory potential of *Rhododendron arboreum* bark. *Toxicol. Ind. Health* 32, 1254–1259.

Nisarshy, M., Ali, S., Qaisar, M., 2013. Antibacterial and cytotoxic activities of the methanolic extracts of *Rhododendron arboreum*. *J. Med. Plants Res.* 7, 398–403.

Nomura, T., Fukai, T., Katayanagi, M., 1978. Studies on the constituents of the cultivated mulberry tree. III. Isolation of four new flavones, kuwanon A, B, C and oxydihydromorusin from the root bark of *Morus alba* L. *Chem. Pharm. Bull.* 26, 1453–1458.

Nomura, T., Fukai, T., Yamada, S., Katayanagi, M., 1976. Phenolic constituents of the cultivated mulberry tree (*Morus alba* L.). *Chem. Pharm. Bull.* 24, 2898–2900.

O. Penelope, *100 Great Natural Remedies*, Kyle Cathic Limited, New York, NY, USA, 1997.

Oèzdemir, M., Açkurt, F., Kaplan, M., Yıldız, M., Löker, M., Gürcan, T., Biringen, G., Okay, A., Seyhan, F.G., 2001. Evaluation of new Turkish hybrid hazelnut (*Corylus avellana* L.) varieties: fatty acid composition, α-tocopherol content, mineral composition and stability. *Food Chem.* 73, 411–415.

Oh, K.-S., Ryu, S.Y., Lee, S., Seo, H.W., Oh, B.K., Kim, Y.S., Lee, B.H., 2009. Melanin-concentrating hormone-1 receptor antagonism and anti-obesity effects of ethanolic extract from *Morus alba* leaves in diet-induced obese mice. *J. Ethnopharmacol.* 122, 216–220.

Oliveira, A.P., Silva, L.R., Andrade, P.B., Valentao, P., Silva, B.M., Goncalves, R.F., Pereira, J.A. and Guedes de Pinho, P., 2010. Further insight into the latex metabolite profile of *Ficus carica*. *Journal of Agricultural and Food Chemistry*, 58(20), pp. 10855-10863.

Oliveira, A.P., Silva, L.R., de Pinho, P.G., Gil-Izquierdo, A., Valentão, P., Silva, B.M., Pereira, J.A. and Andrade, P.B., 2010. Volatile profiling of *Ficus carica* varieties by HS-SPME and GC–IT-MS. *Food Chemistry*, 123(2), pp.548-557.

Oliveira, I., Sousa, A., Valentão, P., Andrade, P.B., Ferreira, I.C.F.R., Ferreres, F., Bento, A., Seabra, R., Estevinho, L., Pereira, J.A., 2007. Hazel (*Corylus avellana* L.) leaves as source of antimicrobial and antioxidative compounds. *Food Chem.* 105, 1018–1025.

References

Oliver, Chadwick Dearing, Larson, B.C., Oliver, C D, 1996. *Forest Stand Dynamics.* Wiley New York.

Ong, H.G., Ling, S.M., Win, T.T.M., Kang, D.-H., Lee, J.-H., Kim, Y.-D., 2018. Ethnobotany of wild medicinal plants used by the Müün ethnic people: A quantitative survey in southern Chin state, Myanmar. *J. Herb. Med.* 13, 91–96.

Orhan, İ.E., Özçelik, B., Şener, B., 2011. Evaluation of antibacterial, antifungal, antiviral, and antioxidant potentials of some edible oils and their fatty acid profiles. *Turkish J. Biol.* 35, 251–258.

Orwa, C., Mutua, A., Kindt, R., Jamnadass, R., Simons, A., 2009. *Agroforestree Database: A Tree Reference and Selection Guide.* Version 4. Agroforestree Database a tree Ref. Sel. Guid. Version 4.

Ottaggio, L., Bestoso, F., Armirotti, A., Balbi, A., Damonte, G., Mazzei, M., Sancandi, M., Miele, M., 2008. Taxanes from shells and leaves of *Corylus avellana.* *J. Nat. Prod.* 71, 58–60.

Painuli, S., Joshi, S., Bhardwaj, A., Meena, R.C., Misra, K., Rai, N., Kumar, N., 2018. In vitro antioxidant and anticancer activities of leaf extracts of *Rhododendron arboreum* and *Rhododendron campanulatum* from Uttarakhand region of India. *Pharmacogn. Mag.* 14, 294.

Painuli, S., Rai, N., Kumar, N., 2016. Gas chromatography and mass spectrometry analysis of methanolic extract of leaves of *Rhododendron arboreum.* *GAS* 9.

Pala, M., Açkurt, F., Löker, M., Yıldız, M., Ömeroğlu, S., 1996. Fındık çeşitlerinin bileşimi ve beslenme fizyolojisi açısından değerlendirilmesi [Evaluation of hazelnut varieties in terms of composition and nutritional physiology]. *Tr. J. Agric. For.* 20, 43–48.

Panda, S., Kirtania, I., 2016. Variation in *Rhododendron arboreum* Sm. complex (Ericaceae): insights from exomorphology, leaf anatomy and pollen morphology. *Mod. Phytomorphology* 9, 27–49.

Panday, K. 1982. Fodder trees and tree fodder in Nepal. *Bern Swiss Dev. Coop. Birmensd. Swiss Fed. Inst. For. Res.*

Pande, P.C., Tiwari, L., Pande, H.C., 2007. *Ethnoveterinary Plants of Uttaranchal—A Review.*

Pandey, N., Kumar, A., Niranjan, A., Lehri, A., Meena, B., Rana, T.S., Tewari, S.K., and Upreti, D.K. 2016. Biochemical composition of *Betula utilis* D. Don bark, collected from high altitudes of Indian Himalayas. *Medicinal Plants-International Journal of Phytomedicines and Related Industries,* 8, 33–39.

Pandey, S., Yadama, G.N., 1990. Conditions for local level community forestry action a theoretical explanation. *Mt. Res. Dev.* 88–95.

Park, S., Han, J., Im, K., Whang, W.K. and Min, H., 2013. Antioxidative and anti-inflammatory activities of an ethanol extract from fig (*Ficus carica*) branches. *Food Science and Biotechnology,* 22(4), pp.1071-1075.

Patil, V.V. and Patil, V.R., 2011. *Ficus carica* Linn.-an overview. *Research Journal of Medicinal Plant,* 5(3), pp.246-253.

Patnaik, R.K., Patnaik, G.N., 1991. Effects of feeding sal (*Shorea robusta*) leaves on the performance of crossbred Jersey cows under farm conditions in Orissa. *Indian Vet. J.*

Patra, A., Dey, A.K., Kundu, A.B., Saraswathy, A., Purushothaman, K.K., 1992. Shoreaphenol, a polyphenol from *Shorea robusta*. *Phytochemistry* 31, 2561–2562.

Paul, A., Khan, M.L., Arunachalam, A., Arunachalam, K., 2005. Biodiversity and conservation of rhododendrons in Arunachal Pradesh in the Indo-Burma biodiversity hotspot. *Curr. Sci.* 89, 623–634.

Paul, A., Khan, M.L., Das, A.K., 2010. Utilization of rhododendrons by Monpas in western Arunachal Pradesh, India. *J. Am. Rhododendron Soc.* 64, 81–84.

Pel, P., Chae, H.-S., Nhoek, P., Kim, Y.-M., Chin, Y.-W., 2017. Chemical constituents with proprotein convertase subtilisin/kexin type 9 mRNA expression inhibitory activity from dried immature *Morus alba* fruits. *J. Agric. Food Chem.* 65, 5316–5321.

Pelvan, E., Alasalvar, C., Uzman, S., 2012. Effects of roasting on the antioxidant status and phenolic profiles of commercial Turkish hazelnut varieties (*Corylus avellana* L.). *J. Agric. Food Chem.* 60, 1218–1223.

Pelvan, E., Olgun, E.Ö., Karadağ, A., Alasalvar, C., 2018. Phenolic profiles and antioxidant activity of Turkish Tombul hazelnut samples (natural, roasted, and roasted hazelnut skin). *Food Chem.* 244, 102–108.

Pérez, C., Canal, J.R., Campillo, J.E., Romero, A. and Torres, M.D., 1999. Hypotriglyceridaemic activity of *Ficus carica* leaves in experimental hypertriglyceridaemic rats. *Phytotherapy Research,* 13(3), pp.188-191.

Pérez-Armada, L., Rivas, S., González, B., Moure, A., 2019. Extraction of phenolic compounds from hazelnut shells by green processes. *J. Food Eng.* 255, 1–8.

Piao, S., Qiu, F., Chen, L., Pan, Y., Dou, D., 2009. New stilbene, benzofuran, and coumarin glycosides from *Morus alba*. *Helv. Chim. Acta* 92, 579–587.

Plants, I.M. and Khare, B.C., 2007. *An Illustrated Dictionary.* Springer Science+ Business Media, LLC., Spring Street, New York, NY USA, pp.1-739.

Pokharel, R.K., 2000. From practice to policy, squatters as forest protectors in Nepal, an experience from Shrijana forest user group. *For. Trees People Newsl.* 31–35.

Pokharel, R.K., Adhikari, S.N., Thapa, Y.B., 1999. Sankarnagar forest user group: learning from a successful FUG in the Tarai. *For. Trees People Newsl.* 25–32.

Pokhriyal, T.C., Ramola, B.C., Raturi, A.S., 1987. Soil moisture regime and nitrogen content in natural sal forest (*Shorea robusta*). *Indian For.* 113, 300–306.

Pothinuch, P., Tongchitpakdee, S., 2019. Phenolic Analysis for Classification of Mulberry (*Morus* spp.) Leaves according to Cultivar and Leaf Age. *J. Food Qual.* 2019.

Poudyal, A.S., 2000. Wildlife corridor management: analysis of biodiversity and socioeconomics in the buffer zone of the Royal Chitawan National Park, Nepal. *Asian Inst. Technol.* Bangkok.

Pradhan, U.C., Lachungpa, S.T., 1990. *Sikkim-Himalayan Rhododendrons.* Primulaceae Books.

Prajapati ND, Purohit SS, Sharma AK, Kumar T. (2007). *A Handbook of Medicinal Plants:* Section II. Jodhpur, India: Agrobios Publishers.

Prakash, D., Upadhyay, G., Singh, B.N., Dhakarey, R., Kumar, S., Singh, K.K., 2007. Free-radical scavenging activities of Himalayan rhododendrons. *Curr. Sci.* 526–532.

Prakash, E.O., Rao, J.T., 1999. A new flavone glycoside from the seeds of *Shorea robusta*. *Fitoterapia* 70, 539–541.

Prakash, T., Fadadu, S.D., Sharma, U.R., Surendra, V., Goli, D., Stamina, P., Kotresha, D., 2008. Hepatoprotective activity of leaves of *Rhododendron arboreum* in CCl4 induced hepatotoxicity in rats. *J. Med. Plants Res.* 2, 315–320.

Prakash, V., Rana, S., Sagar, A., 2016. Studies on antibacterial activity of leaf extracts of *Rhododendron arboreum* and Rhododendron campanulatum. *Int J Curr Microbiol Appl Sci* 5, 315–322.

Prasad, P.V., Subhaktha, P.K., Narayana, A. and Rao, M.M., 2006. Medico-historical study of "aśvattha"(sacred fig tree). *Bulletin of the Indian Institute of History of Medicine (Hyderabad)*, 36(1), pp.1-20.

Prasad, R, Pandey, R.K., 1987. Vegetation damage by frost in natural forests of Madhya Pradesh. *J. Trop. For.*

Prasad, Ram, Pandey, R.K., 1987. Survey of medicinal wealth of Central India. I. Potential of Indigenous medicinal plants in natural forest of eastern Madhya Pradesh. *J. Trop. For.*

Prosperini, S., Ghirardello, D., Scursatone, B., Gerbi, V., Zeppa, G., 2009. Identification of soluble phenolic acids in hazelnut (*Corylus avellana* L.) kernel. *Acta Hortic.* 677–680.

Pullaiah, T., Rani, S.S., 1999. *Trees of Andhra Pradesh*, India. Daya Books.

Qin, C., Li, Y., Niu, W., Ding, Y., Zhang, R., Shang, X., 2010. Analysis and characterisation of anthocyanins in mulberry fruit. *Czech J. Food Sci.* 28, 117–126.

Qureshi SJ, Khan MA, Ahmad M. (2008). A survey of useful medicinal plants of Abbottabad in Northern Pakistan. *Trakia J Sci* 6:39–51.

Rahman, I.U., Afzal, A., Iqbal, Z., Ijaz, F., Ali, N., Bussmann, R.W., 2018. Traditional and ethnomedicinal dermatology practices in Pakistan. *Clin. Dermatol.* 36, 310–319.

Rahmani, A.H. and Aldebasi, Y.H., 2017. *Ficus carica* and its constituents role in management of diseases. *Asian J Pharm Clin Res*, 10(6), pp.49-53.

Rai, T., Rai, L., 1994. *Trees of the Sikkim Himalaya*. Indus Publishing.

Rajan, R.P., 1995. Sal leaf plate processing and marketing in West Bengal. *Soc. Non-Timber For. Prod. Trop. Asia* 27–36.

Ramazani, A., Zakeri, S., Sardari, S., Khodakarim, N. and Djadidt, N.D., 2010. In vitro and in vivo anti-malarial activity of *Boerhavia elegans* and *Solanum surattense*. *Malaria Journal*, 9(1), p.124.

Rana, B.S., Singh, S.P., Singh, R.P., 1988. *Biomass and Productivity of Central Himalayan Sal (*Shorea robusta*) forest*.

Rangaswami, S., Sambamurthy, K., 1959. Chemical examination of the leaves ofRhododendron nilagiricum Zenk, in: *Proceedings of the Indian Academy of Sciences-Section A*. Springer, pp. 366–373.

Rangaswami, S., Sambamurthy, K., 1960. Crystalline chemical components of the flowers of *Rhododendron nilagiricum* Zenk, in: *Proceedings of the Indian Academy of Sciences-Section A*. Springer, pp. 322–327.

Rao, A.R., Singh, B.P., 1996. Non-wood forest products contribution in tribal economy.(A case study in south Bihar and south West Bengal). *Indian For.* 122, 337–342.

Rao, A.V., Snyder, D.M., 2010. Raspberries and human health: a review. *J. Agric. Food Chem.* 58, 3871–3883.

Rashid, K.I., Mahdi, N.M., Alwan, M.A. and Khalid, L.B., 2014. Antimicrobial activity of fig (*Ficus carica* Linn.) leaf extract as compared with latex extract against selected bacteria and fungi. *Journal of University of Babylon*, 22(5), pp.1620-1626.

Ratna Manjula, R., Rao, J.K. and Reddi, T.S., 2011. Ethnomedicinal plants used to cure jaundice in Kammam district of Andhra Pradesh, India. *Journal of Phytology*.

Ravindra, K., Singh, A.K., Abbas, S.G., 1994. Change in population structure of some dominant tree species of dry peninsular sal forest. *Indian For.* 120, 343–348.

Rawat, P., Rai, N., Kumar, N., Bachheti, R.K., 2017. Review on *Rhododendron arboreum*-a magical tree. *Orient. Pharm. Exp. Med.* 17, 297–308.

Raza, R., Ilyas, Z., Sajid, A., Nisar, M., Khokhar, M.Y., Iqbal, J., 2015. Identification of Highly Potent and Selective [alpha]-Glucosidase Inhibitors with Antiglycation Potential, Isolated from *Rhododendron arboreum*. *Rec. Nat. Prod.* 9, 262.

Rhododendron, S., 1980. *Text Book of Materia Medica*.

Rieger, M., 2006. *Introduction to Fruit Crops*. CRC Press.

Riethmüller, E., Alberti, Á., Tóth, G., Béni, S., Ortolano, F., Kéry, Á., 2013. Characterisation of diarylheptanoid-and flavonoid-type phenolics in *Corylus avellana* L. leaves and bark by HPLC/DAD–ESI/MS. *Phytochem. Anal.* 24, 493–503.

Rivière, C., Krisa, S., Péchamat, L., Nassra, M., Delaunay, J.-C., Marchal, A., Badoc, A., Waffo-Téguo, P., Mérillon, J.-M., 2014. Polyphenols from the stems of *Morus alba* and their inhibitory activity against nitric oxide production by lipopolysaccharide-activated microglia. *Fitoterapia* 97, 253–260.

Rohela, G.K., Muttanna, P.S., Kumar, R., Chowdhury, S.R., 2020. Mulberry (Morus spp.): An Ideal Plant for Sustainable Development. *Trees, For. People 100011*.

Rolim, P.M., 2015. Development of prebiotic food products and health benefits. *Food Sci. Technol.* 35, 3–10.

Rønsted, N., Weiblen, G.D., Savolainen, V. and Cook, J.M., 2008. Phylogeny, biogeography, and ecology of Ficus section Malvanthera (Moraceae). *Molecular Phylogenetics and Evolution*, 48(1), pp.12-22.

Rostam, M.A., Salleh, M.S., Harun, A.F. and Ali, A.N., 2018. The Antimicrobial Activities of Fig (*Ficus carica* L.) Leaves Extract Against Staphylococcus aureus and Escherechia coli. *International Journal of Allied Health Sciences*, 2(1), pp.273-284.

Rothwell, J.A., Perez-Jimenez, J., Neveu, V., Medina-Remon, A., M'Hiri, N., García-Lobato, P., Manach, C., Knox, C., Eisner, R., Wishart, D.S., 2013. *Phenol-Explorer 3.0: A Major update of the Phenol-Explorer Database to Incorporate Data on the Effects of Food Processing on Polyphenol Content*. Database 2013.

Roy, J.D., Handique, A.K., Barua, C.C., Talukdar, A., Ahmed, F.A., Barua, I.C., 2014. Evaluation of phytoconstituents and assessment of adaptogenic activity in vivo in various extracts of *Rhododendron arboreum* (leaves). *Indian J. Pharm. Biol. Res.* 2, 49–56.

Rubnov, S., Kashman, Y., Rabinowitz, R., Schlesinger, M. and Mechoulam, R., 2001. Suppressors of cancer cell proliferation from fig (*Ficus carica*) resin: isolation and structure elucidation. *Journal of Natural Products*, 64(7), pp.993-996.

Saeed, M.A. and Sabir, A.W., 2002. Irritant potential of triterpenoids from *Ficus carica* leaves. *Fitoterapia,* 73(5), pp.417-420.

Sah, S.P., 1996. Natural regeneration status of sal (*Shorea robusta* Gaertn.) forests in Nepal, in: *Modelling Regeneration Success and Early Growth of Forest Stands, Copenhagen* (Denmark), 10-13 Jun 1996. FSL.

Sakai, A., and Larcher, W. 2012. *Frost Survival of Plants: Responses and Adaptation to Freezing Stress.* Springer Science & Business Media.

Saklani, S., Chandra, S., 2015. Evaluation of in vitro antimicrobial activity, nutritional profile and phytochemical screening of *Rhododendron arboreum. World J Pharm Sci* 4, 962–971.

Samant, Pant, S., Singh, M., Lal, M., Singh, A., Sharma, A., and Bhandari, S. 2007. Medicinal plants in Himachal Pradesh, north western Himalaya, India. Int. J. Biodivers. Sci. Manag. *The International Journal of Biodiversity Science and Management,* 3, 234–251.

Sanchez, M.D., 2002. Mulberry for Animal Production: *Proceedings of an Electronic Conference Carried out between May and August 2000.* Food & Agriculture Org.

Sánchez-Salcedo, E.M., Mena, P., García-Viguera, C., Hernández, F., Martínez, J.J., 2015a. (Poly) phenolic compounds and antioxidant activity of white (*Morus alba*) and black (*Morus nigra*) mulberry leaves: Their potential for new products rich in phytochemicals. *J. Funct. Foods* 18, 1039–1046.

Sánchez-Salcedo, E.M., Mena, P., García-Viguera, C., Martínez, J.J., Hernández, F., 2015b. Phytochemical evaluation of white (*Morus alba* L.) and black (*Morus nigra* L.) mulberry fruits, a starting point for the assessment of their beneficial properties. *J. Funct. Foods* 12, 399–408.

Sánchez-Salcedo, E.M., Sendra, E., Carbonell-Barrachina, Á.A., Martínez, J.J., Hernández, F., 2016. Fatty acids composition of Spanish black (*Morus nigra* L.) and white (*Morus alba* L.) mulberries. *Food Chem.* 190, 566–571.

Santhoshkumar, M., Anusuya, N., Bhuvaneswari, P., 2012. Antiulcerogenic effect of resin from *Shorea robusta* Gaertn. f. on experimentally induced ulcer models. *Int J Pharm Pharm Sci* 5, 269–272.

Saoudi, M. and El Feki, A., 2012. Protective role of *Ficus carica* stem extract against hepatic oxidative damage induced by methanol in male Wistar rats. *Evidence-Based Complementary and Alternative Medicine,* 2012.

Saranya, D., Ravi, R., 2016. The leaf of Nilgiri Rhododendron: a potent antimicrobial agent against medically critical human pathogens. *Therapy* 4, 5.

Saranya, D., Ravi, R., 2017. Nilgiri Rhododendron: a high altitude medicinal tree explored for its antimicrobial activity. *Int. J. Pharm. Sci. Res.* 8, 1830.

Savill, P.S., 2019. *The Silviculture of Trees Used in British Forestry.* CABI.

Sawarkar, H.A., Singh, M.K., Pandey, A.K. and Biswas, D. 2011. In vitro anthelmintic activity of *Ficus bengalhensis, Ficus caria & Ficus religiosa*: a comparative anthelmintic activity. *International J PharmTech Research,* 3, pp.152-153.

Saxena, K.G., Rao, K.S., Purohit, A.N., 1993. Sustainable forestry-prospects in India. *J. Sustain. For.* 1, 69–95.

Schmidt, M.G., Schreier, H., Shah, P.B., 1993. Factors affecting the nutrient status of forest sites in a mountain watershed in Nepal. *J. Soil Sci.* 44, 417–425.

Shafiuddin, M., Abdullah, K., Sadath, A., 2009. Wound healing activity of traditional herbal formulation. *Int. J. Chem. Sci.* 7, 639–643.

Shahidi, F., Alasalvar, C., Liyana-Pathirana, C.M., 2007. Antioxidant phytochemicals in hazelnut kernel (*Corylus avellana* L.) and hazelnut byproducts. *J. Agric. Food Chem.* 55, 1212–1220.

Shahinuzzaman, M., Yaakob, Z., Anuar, F.H., Akhtar, P., Kadir, N.H.A., Hasan, A.M., Sobayel, K., Nour, M., Sindi, H., Amin, N. and Sopian, K., 2020. In vitro antioxidant activity of *Ficus carica* L. latex from 18 different cultivars. *Scientific Reports*, 10(1), pp.1-14.

Shakya, C.M., Bhattarai, D.R., 1995. Market survey of non-wood forest products in Bara and Rautahat districts for operational forest management plans. *For. Manag. Util. Dev. Prot. Dep. For.* Kathmandu, Nepal.

Shanmugam, S., Annadurai, M., Rajendran, K., 2011. Ethnomedicinal plants used to cure diarrhoea and dysentery in Pachalur hills of Dindigul district in Tamil Nadu, Southern India. *J. Appl. Pharm. Sci.* 1, 94.

Sharma, B.C., 2013. In vitro antibacterial activity of certain folk medicinal plants from Darjeeling Himalayas used to treat microbial infections. *J. Pharmacogn. Phytochem.* 2, 1–4.

Sharma, B.K., 1981. Further Studies on Seed Production in Sal (*Shorea robusta* Gaertn.) Crops in Dehra Dun District (UP). *Indian For.* 107, 505–509.

Sharma, N., Sharma, U.K., Gupta, A.P., Sinha, A.K., 2010. Simultaneous determination of epicatechin, syringic acid, quercetin-3-O-galactoside and quercitrin in the leaves of Rhododendron species by using a validated HPTLC method. *J. Food Compos. Anal.* 23, 214–219.

Sharma, P.P., Roy, B.R.K., and Dinesh, G. 2010. Pentacyclic triterpinoids from *Betula utilis* and Hyptis suaveolens. *International Journal of PharmTech Research*, 2, 1558–1562.

Sharma, T.C., Sharma, A.K., Sharma, K.K., Payal, P., Sharma, M.C., Dobhal, M.P., 2014. Phytoconstituents: Isolation and Characterization from Root Bark of *Shorea robusta* Plant. *Int. J. Pharmacogn. Pharm. Res.* 6, 991–995.

Sharma, U.R., Surendra, V., Jha, S.K., Nitesh, S.C., Prakash, T., Goli, D., 2009. Evaluation of anti-inflammatory activity of *Rhododendron arboreum* herb extract on experimental animal. *Arch Pharm Sci Res* 1, 58–61.

Shataer, D., Abdulla, R., Ma, Q.L., Liu, G.Y., Aisa, H.A., 2020. Chemical Composition of Extract of *Corylus avellana* Shells. *Chem. Nat. Compd.* 1–3.

Shengelia, Z., 1983. *The Culture of Medicinal Plants in Georgia*. Tbilisi Sabchota Sakartv.

Sher, Z., Khan, Z. and Hussain, F., 2011. Ethnobotanical studies of some plants of Chagharzai valley, district Buner, Pakistan. *Pak J Bot*, 43(3), pp.1445-1452.

Shilajan, S., Swar, G., 2013. Simultaneous estimation of three triterpenoids-ursolic acid, b-sitosterol and lupeol from flowers, leaves and formulations of *Rhododendron arboreum* Smith. using validated HPTLC method. *Int. J. Green Pharm.* 7.

Šiman, P., Filipová, A., Tichá, A., Niang, M., Bezrouk, A., and Havelek, R. 2016. Effective method of purification of betulin from birch bark: the importance of its purity for scientific and medicinal use. *PLoS One*, 11, e0154933.

Simsek, A., Aykut, O., 2007. Evaluation of the microelement profile of Turkish hazelnut (*Corylus avellana* L.) varieties for human nutrition and health. *Int. J. Food Sci. Nutr.* 58, 677–688.

Singab, A.N.B., El-Beshbishy, H.A., Yonekawa, M., Nomura, T., Fukai, T., 2005. Hypoglycemic effect of Egyptian *Morus alba* root bark extract: effect on diabetes and lipid peroxidation of streptozotocin-induced diabetic rats. *J. Ethnopharmacol.* 100, 333–338.

Singh, D., Singh, B. and Goel, R.K., 2011. Traditional uses, phytochemistry and pharmacology of *Ficus religiosa*: A review. *Journal of Ethnopharmacology*, 134(3), pp.565-583.

Singh, O., Sharma, D.C., Rawat, J.K., 1993. Production and decomposition of leaf litter in sal, teak, eucalyptus and poplar forests in Uttar Pradesh. *Indian For.* 119, 112–121.

Singh, R.P., Chaturvedi, O.P., 1983. Primary production of a deciduous forest at Varanasi. *Indian For.* 109, 255–260.

Sinha, R.P., Nath, K., 1982. Effect of urea supplementation on nutritive value of deoiled sal-meal in cattle. *Indian J. Anim. Sci.*

Slatnar, A., Klancar, U., Stampar, F. and Veberic, R., 2011. Effect of drying of figs (*Ficus carica* L.) on the contents of sugars, organic acids, and phenolic compounds. *Journal of Agricultural and Food Chemistry*, 59(21), pp.11696-11702.

Slatnar, A., Mikulic-Petkovsek, M., Stampar, F., Veberic, R., Solar, A., 2014. HPLC-MSn identification and quantification of phenolic compounds in hazelnut kernels, oil and bagasse pellets. *Food Res. Int.* 64, 783–789.

Soetan, K.O., Olaiya, C.O., Oyewole, O.E., 2010. The importance of mineral elements for humans, domestic animals and plants-A review. *African J. food Sci.* 4, 200–222.

Sohn, H.-Y., Son, K.H., Kwon, C.-S., Kwon, G.-S., Kang, S.S., 2004. Antimicrobial and cytotoxic activity of 18 prenylated flavonoids isolated from medicinal plants: *Morus alba* L., *Morus mongolica* Schneider, *Broussnetia papyrifera* (L.) Vent, *Sophora flavescens* Ait and *Echinosophora koreensis* Nakai. *Phytomedicine* 11, 666–672.

Solar, A., Stampar, F., 2011. Characterisation of selected hazelnut cultivars: phenology, growing and yielding capacity, market quality and nutraceutical value. *J. Sci. Food Agric.* 91, 1205–1212.

Solar, A., Veberič, R., Bacchetta, L., Botta, R., Drogoudi, P., Metzidakis, I., Rovira, M., Sarraquigne, J.P., Silva, A.P., 2008. Phenolic characterization of some hazelnut cultivars from different European germplasm collections, in: *VII International Congress on Hazelnut* 845. pp. 613–618.

Solomon, A., Golubowicz, S., Yablowicz, Z., Grossman, S., Bergman, M., Gottlieb, H.E., Altman, A., Kerem, Z. and Flaishman, M.A., 2006. Antioxidant activities and anthocyanin content of fresh fruits of common fig (*Ficus carica* L.). *Journal of Agricultural and Food Chemistry*, 54(20), pp.7717-7723.

Sonar, P.K., Singh, R., Bansal, P., Balapure, A.K., Saraf, S.K., 2012a. *R. arboreum* flower and leaf extracts: RP-HPTLC screening, isolation, characterization and biological activity. *Rasayan J Chem* 5, 165–172.

Sonar, P.K., Singh, R., Khan, S., Saraf, S.K., 2012b. Isolation, characterization and activity of the flowers of *Rhododendron arboreum* (Ericaceae). *J. Chem.* 9, 631–636.

Sonar, P.K., Singh, R., Verma, A., Saraf, S.K., 2013. *Rhododendron arboreum* (Ericaceae): Immunomodulatory and related toxicity studies. *Orient. Pharm. Exp. Med.* 13, 127–131.

Soonthornsit, N., Pitaksutheepong, C., Hemstapat, W., Utaisincharoen, P., Pitaksuteepong, T., 2017. In vitro anti-inflammatory activity of *Morus alba* L. stem extract in LPS-stimulated RAW 264.7 cells. *Evidence-Based Complementary and Alternative Medicine.* 2017.

Souci, S.W., Fachmann, W., Kraut, H., 2000. *Food Composition and Nutrition Tables.* Medpharm GmbH Scientific Publishers.

Srivastava, P., 2012. *Rhododendron arboreum*: an overview. *J. Appl. Pharm. Sci.* 2, 158–162.

Stalin, C., Dineshkumar, P. and Nithiyananthan, K., 2012. Evaluation of antidiabetic activity of methanolic leaf extract of *Ficus carica* in alloxan-induced diabetic rats. *Asian J Pharm Clin Res,* 5(3), pp.85-7.

Stephen Irudayaraj, S., Christudas, S., Antony, S., Duraipandiyan, V., Naif Abdullah, A.D. and Ignacimuthu, S., 2017. Protective effects of *Ficus carica* leaves on glucose and lipids levels, carbohydrate metabolism enzymes and β-cells in type 2 diabetic rats. *Pharmaceutical biology,* 55(1), pp.1074-1081.

Strack, D., Meurer, B., Wray, V., Grotjahn, L., Austenfeld, F.A., Wiermann, R., 1984. Quercetin 3-glucosylgalactoside from pollen of *Corylus avellana. Phytochemistry* 23, 2970–2971.

Sundriyal, R.C., Sharma, E., Rai, L.K., Rai, S.C., 1994. Tree structure, regeneration and woody biomass removal in a sub-tropical forest of Mamlay watershed in the Sikkim Himalaya. *Vegetatio* 113, 53–63.

Supriya, K., Kotagiri, S., Swamy, V.B.M., Swamy, A.P., Vishwanath, K.M., 2012. Anti-Obesity activity of *Shorea robusta* G. leaves extract on monosodium glutamate induced obesity in albino rats. *Res. J. Pharm. Biol. Chem. Sci.* 3, 555–565.

Suresh, C., 1994. Rural development through forest-based cottage industries. *Vaniki Sandesh* 18, 1–5.

Swaroop, A., Gupta, A.P., Sinha, A.K., 2005. Simultaneous determination of quercetin, rutin and coumaric acid in flowers of *Rhododendron arboreum* by HPTLC. *Chromatographia* 62, 649.

Tambe, V.D., Katariya, T.R., Jadhav, R.S. and Singh, S.D., 2009. Anthelmintic Activity of Bark of Ficus Carica. *Research Journal of Pharmacy and Technology,* 2(1), pp.206-207.

Tene, V., Malagon, O., Finzi, P.V., Vidari, G., Armijos, C. and Zaragoza, T., 2007. An ethnobotanical survey of medicinal plants used in Loja and Zamora-Chinchipe, Ecuador. *Journal of Ethnopharmacology,* 111(1), pp.63-81.

Tewari, D.N., 1995. A monograph on Sal (*Shorea robusta* Gaertn. f). *International Book Distributors.*

Thabti, I., Elfalleh, W., Hannachi, H., Ferchichi, A., Campos, M.D.G., 2012. Identification and quantification of phenolic acids and flavonol glycosides in Tunisian Morus species by HPLC-DAD and HPLC–MS. *Journal of Functional Foods* 4, 367–374.

Thacker, P., Gautam, K.H., 1994. *A Socio-Economic Study of Participatory Issues in Forest Management in the Terai.* HMGN/FINNIDA.

Tikader, A., 2011. Distribution, diversity, utilization and conversation of mulberry (*Morus* spp.) in North West of India. *Asian and Australasian Journal of Plant Science and Biotechnology.* 5, 67–72.

Tiwari, O.N., Chauhan, U.K., 2006. Rhododendron conservation in Sikkim Himalaya. *Curr. Sci.* 532–541.

Torello Marinoni, D., Valentini, N., Portis, E., Acquadro, A., Beltramo, C., Mehlenbacher, S.A., Mockler, T.C., Rowley, E.R., Botta, R., 2018. High density SNP mapping and QTL analysis for time of leaf budburst in *Corylus avellana* L. *PLoS One* 13, e0195408.

Tsaturyan, T., Gevorgyan, M., 2007. *Wild Edible Plants of Armenia.* Yerevan Armen. Acad. Sci.

Tsaturyan, T., Gevorgyan, M., 2014. *Wild Medicinal Plants of Armenia.* Yerevan Armen. Acad. Sci.

Turan, G.Y.M., Akccedil, F., 2011. The importance of Turkish hazelnut trace and heavy metal contents for human nutrition. *J. Soil Sci. Environ. Manag.* 2, 25–33.

Turova, A.D., Sapozhnikova, E.N., 1974. *Medicinal Plants of the USSR and their Use.* Meditsina, Moscow 209.

Ugulu, I., 2011. Traditional ethnobotanical knowledge about medicinal plants used for external therapies in Alasehir, Turkey. *Int. J. Med. Arom. Plants,* 1(2), pp.101-106.

Ugulu, I., Baslar, S., Yorek, N. and Dogan, Y., 2009. The investigation and quantitative ethnobotanical evaluation of medicinal plants used around Izmir province, Turkey. *Journal of Medicinal plants research,* 3(5), pp.345-367.

Uniyal, S.K., Singh, K.N., Jamwal, P., Lal, B., 2006. Traditional use of medicinal plants among the tribal communities of Chhota Bhangal, Western Himalaya. *J. Ethnobiol. Ethnomed.* 2, 14.

Upadhyay, L.R., 1992. Use of tree fodder in Jhapa and Sunsari Districts in the eastern Terai. *Banko Janakari* 3, 17–18.

Urban, M., Sarek, J., Klinot, J., Korinkova, G., and Hajduch, M. 2004. Synthesis of A-seco derivatives of betulinic acid with cytotoxic activity. *Journal of Natural Products,* 67, 1100–1105.

Ustaoglu, B., Karaca, M., 2014. The effects of climate change on spatiotemporal changes of hazelnut (*Corylus avellana*) cultivation areas in the Black Sea region, Turkey. *Appl. Ecol. Environ. Res.* 12, 309–324.

Uysal, İ., Onar, S., Karabacak, E. and Çelik, S., 2010. Ethnobotanical aspects of Kapıdağ Peninsula (Turkey). *Biological Diversity and Conservation,* 3(3), pp.15-22.

Vallejo, F., Marín, J.G. and Tomás-Barberán, F.A., 2012. Phenolic compound content of fresh and dried figs (*Ficus carica* L.). *Food Chemistry,* 130(3), pp.485-492.

Vashisht, S., Singh, M.P., Chawla, V., 2016. In-vitro antioxidant and antibacterial activity of methanolic extract of *Shorea robusta* Gaertn. F. resin. *Int. J. Pharm. Phytopharm. Res.* 6, 68–71.

Vaya, J. and Mahmood, S., 2006. Flavonoid content in leaf extracts of the fig (*Ficus carica* L.), carob (*Ceratonia siliqua* L.) and pistachio (*Pistacia lentiscus* L.). *Biofactors,* 28(3-4), pp.169-175.

Veberic, R., Colaric, M. and Stampar, F., 2008. Phenolic acids and flavonoids of fig fruit (*Ficus carica* L.) in the northern Mediterranean region. *Food Chemistry*, 106(1), pp.153-157.

Veberic, R., Jakopic, J. and Stampar, F., 2008. Internal fruit quality of figs (*Ficus carica* L.) in the Northern Mediterranean Region. *Italian Journal of Food Science*, 20(2), pp.255-262.

Venkat, V.P.S., Sharma, B.K., 1978. Studies on production and collection of sal (*Shorea robusta* Gaertn.) seeds. *Indian For.* 104, 414–420.

Venkateswaran, V., Fleshner, N.E., Klotz, L.H., 2002. Modulation of cell proliferation and cell cycle regulators by vitamin E in human prostate carcinoma cell lines. *J. Urol.* 168, 1578–1582.

Verma, N., Amresh, G., Sahu, P.K., Mishra, N., Rao, C. V, Singh, A.P., 2013. Pharmacological Evaluation of Hyperin for Antihyperglycemic Activity and Effect on Lipid Profile in Diabetic Rats. *Indian J Exp Biol.* 2013 Jan;51(1):65-72.

Verma, N., Amresh, G., Sahu, P.K., Rao, C. V, Singh, A.P., 2012. Antihyperglycemic and antihyperlipidemic activity of ethyl acetate fraction of *Rhododendron arboreum* Smith flowers in streptozotocin induced diabetic rats and its role in regulating carbohydrate metabolism. *Asian Pac. J. Trop. Biomed.* 2, 696.

Verma, N., Singh, A.P., Amresh, G., Sahu, P.K., Rao, C. V, 2011a. Protective effect of ethyl acetate fraction of *Rhododendron arboreum* flowers against carbon tetrachloride-induced hepatotoxicity in experimental models. *Indian J. Pharmacol.* 43, 291.

Verma, N., Singh, A.P., Amresh, G., Sahu, P.K., Rao, C., 2010. Anti-inflammatory and anti-nociceptive activity of *Rhododendron arboreum*. *J Pharm Res* 3, 1376–1380.

Verma, N., Singh, A.P., Gupta, A., Sahu, P.K., Rao, C. V, 2011b. Antidiarrheal potential of standardized extract of *Rhododendron arboreum* Smith flowers in experimental animals. *Indian J. Pharmacol.* 43, 689.

Vinson, J.A., 1999. The functional food properties of figs. *Cereal Foods World*, 44(2), pp.82-87.

Vu, C.C., Verstegen, M.W.A., Hendriks, W.H., Pham, K.C., 2011. The nutritive value of mulberry leaves (*Morus alba*) and partial replacement of cotton seed in rations on the performance of growing Vietnamese cattle. *Asian-Australasian Journal of Animal Sciences.* 24, 1233–1242.

Wagner, W.L. and Lorence, D., 1990. *Manual of the Flowering Plants of Hawaii.* 2 vols. Vol. 1, Rubiaceae. Honolulu.

Wagner, W.L., Herbst, D.R. and Sohmer, S.H., 1990. *Manual of the Flowering Plants of Hawaii.* University of Hawaii Press.

Wang, G., Wang, H., Song, Y., Jia, C., Wang, Z. and Xu, H., 2004. Studies on anti-HSV effect of *Ficus carica* leaves. *Journal of Chinese Medicinal Materials*, 27(10), pp.754-756.

Wang, J., Wang, X., Jiang, S., Lin, P., Zhang, J., Lu, Y., Wang, Q., Xiong, Z., Wu, Y., Ren, J. and Yang, H., 2008. Cytotoxicity of fig fruit latex against human cancer cells. *Food and Chemical Toxicology*, 46(3), pp.1025-1033.

Wang, W., Zu, Y., Fu, Y., Efferth, T., 2012. In vitro antioxidant and antimicrobial activity of extracts from *Morus alba* L. leaves, stems and fruits. *The American Journal of Chinese Medicine.* 40, 349–356.

Wang, Y., Xiang, L., Wang, C., Tang, C., He, X., 2013. Antidiabetic and antioxidant effects and phytochemicals of mulberry fruit (*Morus alba* L.) polyphenol enhanced extract. *PLoS One* 8, e71144.

Wani, M.S., Gupta, R.C., Munshi, A.H., and Pradhan, S.K. 2018a. Phytochemical screening, total phenolics, flavonoid content and antioxidant potential of different parts of *Betula utilis* D. Don from Kashmir Himalaya. *International Journal of Pharmaceutical Sciences and Research*, 9, 2411–2417.

Wani, M.S., Gupta, R.C., Pradhan, S.K., and Munshi, A.H. 2018b. Estimation of four triterpenoids, betulin, lupeol, oleanolic acid, and betulinic acid, from bark, leaves, and roots of *Betula utilis* D. Don using a validated high-performance thin-layer chromatographic method. *JPC-Journal Planar Chromatogr. TLC*, 31, 220–229.

Wani, M.S., Tantary, Y.R., Gupta, R.C., Jan, I., and Zarger, S.A. 2020. Betula L.: Distribution, Ecology and Phytochemicals. In *Betula: Ecology and Uses*, edited by Carl T. Bertelsen Nova Publisher.

Wani, T A, Chandrashekara, H.H., Kumar, D., Prasad, R., Gopal, A., Sardar, K.K., Tandan, S.K., 2012a. Wound Healing Activity of Ethanolic Extract of *Shorea robusta* Gaertn. f. resin. *Indian J Biochem Biophys.* 2012 Dec;49(6):463-7.

Wani, T A, Chandrashekara, H.H., Kumar, D., Prasad, R., Sardar, K.K., Tandan, S.K., 2012b. Anti-inflammatory and Antipyretic Activities of the Ethanolic Extract of *Shorea robusta* Gaertn. f. resin. *Indian J Biochem Biophys.* 2012 Dec;49(6):463-7.

Wani, Tariq Ahmad, Kumar, Dhirendra, Prasad, R., Verma, P.K., Sardar, K.K., Tandan, S.K., Kumar, Dinesh, 2012. Analgesic activity of the ethanolic extract of *Shorea robusta* resin in experimental animals. *Indian J. Pharmacol.* 44, 493.

Watson, L., Dallwitz, M.J., 2007. Moraceae. Fam. Flower. Plants Descr. Illus. Identification, *Inf. Retrieval,* 1992 onward.

Wattanapitayakul, S.K., Chularojmontri, L., Herunsalee, A., Charuchongkolwongse, S., Niumsakul, S., Bauer, J.A., 2005. Screening of antioxidants from medicinal plants for cardioprotective effect against doxorubicin toxicity. *Basic & Clinical Pharmacology & Toxicology.* 96, 80–87.

Webb, E.L., Sah, R.N., 2003. Structure and diversity of natural and managed sal (*Shorea robusta* Gaertn. f.) forest in the Terai of Nepal. *For. Ecol. Manage.* 176, 337–353.

Weli, A.M., Al-Blushi, A.A.M. and Hossain, M.A., 2015. Evaluation of antioxidant and antimicrobial potential of different leaves crude extracts of Omani *Ficus carica* against food borne pathogenic bacteria. *Asian Pacific Journal of Tropical Disease*, 5(1), pp.13-16.

Werbach, M.R., 1993. *Healing through Nutrition: A Natural Approach to Treating 50 Common Illnesses with Diet and Nutrients.* Harpercollins.

Whelan, J., 2008. The Health Implications of Prostaglandins, Leukotrienes and Essential Fatty Acids 79, 165–167.

Wongwat, T., Srihaphon, K., Pitaksutheepong, C., Boonyo, W., Pitaksuteepong, T., 2020. Suppression of inflammatory mediators and matrix metalloproteinase (MMP)-13 by

Morus alba stem extract and oxyresveratrol in RAW 264.7 cells and C28/I2 human chondrocytes. *Journal of Traditional and Complementary Medicine.* 10, 132–140.

Woodland DW. (1997). *Contemporary Plant Systematics,* 2nd ed. Berrien Springs (MI): Andrews University Press.

Woodland, D.W., 1991. *Contemporary Plant Systematics.* Prentice Hall.

World Health Organization, 1985. Energy and protein requirements: report of a joint FAO/WHO/UNU expert consultation. In E*nergy and Protein Requirements: Report of a Joint FAO/WHO/UNU Expert Consultation* (pp. 206-206).

Yadav, A. V, Kawale, L.A., Nade, V.S., 2008. Effect of *Morus alba* L.(mulberry) leaves on anxiety in mice. *Indian Journal of Pharmacology.* 40, 32.

Yang, L., Wen, K.-S., Ruan, X., Zhao, Y.-X., Wei, F., Wang, Q., 2018. Response of plant secondary metabolites to environmental factors. *Molecules* 23, 762.

Yang, N.-C., Jhou, K.-Y., Tseng, C.-Y., 2012. Antihypertensive effect of mulberry leaf aqueous extract containing γ-aminobutyric acid in spontaneously hypertensive rats. *Food Chemistry.* 132, 1796–1801.

Yang, X., Yang, L., Zheng, H., 2010. Hypolipidemic and antioxidant effects of mulberry (*Morus alba* L.) fruit in hyperlipidaemia rats. *Food and Chemical Toxicology.* 48, 2374–2379.

Yang, X.M., Yu, W., Ou, Z.P., Liu, W.M. and Ji, X.L., 2009. Antioxidant and immunity activity of water extract and crude polysaccharide from *Ficus carica* L. fruit. *Plant Foods for Human Nutrition,* 64(2), pp.167-173.

Yang, Y., Gong, T., Liu, C., Chen, R.-Y., 2010a. Four new 2-arylbenzofuran derivatives from leaves of *Morus alba* L. *Chemical and Pharmaceutical Bulletin.* 58, 257–260.

Yang, Y., Zhang, T., Xiao, L., Yang, L., Chen, R., 2010b. Two new chalcones from leaves of *Morus alba* L. *Fitoterapia* 81, 614–616.

Yang, Z., Wang, Yingchao, Wang, Yi, Zhang, Y., 2012. Bioassay-guided screening and isolation of α-glucosidase and tyrosinase inhibitors from leaves of *Morus alba. Food Chemistry.* 131, 617–625.

Yang, Z.-G., Matsuzaki, K., Takamatsu, S., Kitanaka, S., 2011. Inhibitory effects of constituents from *Morus alba* var. multicaulis on differentiation of 3T3-L1 cells and nitric oxide production in RAW264. 7 cells. *Molecules* 16, 6010–6022.

Yaseen Khan, M., Ali, S.A., Pundarikakshudu, K., 2016. Wound healing activity of extracts derived from *Shorea robusta* resin. *Pharm. Biol.* 54, 542–548.

Yemiş, O., Bakkalbaşı, E. and Artık, N., 2012. Changes in pigment profile and surface colour of fig (*Ficus carica* L.) during drying. *International Journal of Food Science & Technology,* 47(8), pp.1710-1719.

Yiemwattana, I., Chaisomboon, N., Jamdee, K., 2018. Antibacterial and anti-inflammatory potential of *Morus alba* stem extract. *The Open Dentistry Journal. J.* 12, 265.

Yu, F., Skidmore, A.K., Wang, T., Huang, J., Ma, K. and Groen, T.A., 2017. *Rhododendron* diversity patterns and priority conservation areas in China. *Diversity and Distributions,* 23: 1143-1156.

Yuan, B., Lu, M., Eskridge, K.M., Isom, L.D., Hanna, M.A., 2018. Extraction, identification, and quantification of antioxidant phenolics from hazelnut (*Corylus avellana* L.) shells. *Food Chem.* 244, 7–15.

Zabihullah, Q., Rashid, A. and Akhtar, N., 2006. Ethnobotanical survey in kot Manzaray Baba valley Malakand agency, Pakistan. *Pak J Plant Sci*, 12(2), pp.115-121.

Zaki, M., Sofi, M.S., and Kaloo, Z.A. 2011. A reproducible protocol for raising clonal plants from leaf segments excised from mature trees of *Betula utilis* a threatened tree species of Kashmir Himalayas. *International Multidisciplinary Research Journal*, 1, 07-13

Zeni, A.L.B., Dall'Molin, M., 2010. Hypotriglyceridemic effect of *Morus alba* L., Moraceae, leaves in hyperlipidemic rats. *Revista Brasileira de Farmacognosia*. 20, 130–133.

Zeppa, D.G.P., Gerbi, V., 2010. Phenolic acid profile and antioxidant capacity of hazelnut (*Corylus avellana* L.) kernels in different solvent systems. *J. Food Nutr. Res.* 49, 195–205.

Zha, H.-G., Milne, R.I., Zhou, H.-X., Chen, X.-Y., Sun, H., 2016. Identification and cloning of class II and III chitinases from alkaline floral nectar of *Rhododendron irroratum*, Ericaceae. *Planta* 244, 805–818.

Zhang, M., Rong-Rong, W., Man, C., Zhang, H.-Q., Shi, S.U.N., Zhang, L.-Y., 2009. A new flavanone glycoside with anti-proliferation activity from the root bark of *Morus alba*. *Chinese Journal of Natural Medicines*. 7, 105–107.

Zhang, W., Han, F., He, J., Duan, C., 2008. HPLC-DAD-ESI-MS/MS analysis and antioxidant activities of nonanthocyanin phenolics in mulberry (*Morus alba* L.). *Journal of Food Science*. 73, C512–C518.

Zhao, L., Cao, B., Wang, F., Yazaki, Y., 1994. Chinese wattle tannin adhesives suitable for producing exterior grade plywood in China. *Holz als Roh-und Werkst.* 52, 113–118.

Zhasa, N.N., Hazarika, P., Tripathi, Y.C., 2015. Indigenous knowledge on utilization of plant biodiversity for treatment and cure of diseases of human beings in Nagaland, India, A case study. *Int. Res. J. Biol. Sci.* 4, 89–106.

Zobel, D.B., and Singh, S.P. 1997. Himalayan forests and ecological generalizations. *Bioscience* 47, 735–745.

Zohary, D., Hopf, M. and Weiss, E., 2012. *Domestication of Plants in the Old World: The Origin and Spread of Domesticated Plants in Southwest Asia, Europe, and the Mediterranean Basin*. Oxford University Press on Demand.

Zolotnitskaya, S., 1965. *Medicinal Resources of the Flora of Armenia*, vols. 1–2. Yerevan Armen. Acad. Sci.

Index

A

acetone, 20, 21, 37, 39, 50, 72, 74, 85
acid, 5, 6, 15, 16, 17, 18, 19, 33, 35, 49, 53, 54, 60, 61, 64, 65, 67, 69, 70, 82, 83, 84, 96, 103, 106, 115, 121, 122, 127, 129, 130, 132, 133, 134
Afghanistan, 3, 24, 45
age, 1, 44, 57
alkaloids, 52, 63, 80, 116, 117
allergy, 52, 81, 82
amino acids, 28, 59, 61, 117
amylase, 33, 52, 98, 102
analgesic, 77, 79, 82, 95
antibacterial, 7, 46, 47, 85, 91, 95, 101, 102, 103, 106, 108, 117, 118, 121, 122, 124, 127, 130, 133
anticancer, 5, 15, 29, 34, 47, 48, 75, 82, 106, 109, 119, 122
antidiabetic, 7, 34, 40, 48, 49, 63, 71, 75, 77, 81, 87, 88, 102, 104, 110, 114, 119, 129, 132
antifungal, 14, 34, 47, 86, 91, 95, 103, 115, 122
anti-inflammatory, 5, 7, 15, 28, 29, 37, 51, 63, 71, 75, 77, 81, 87, 91, 95, 101, 102, 109, 121, 122, 127, 129, 131, 132, 133
antioxidant, 7, 13, 15, 20, 22, 23, 29, 34, 37, 42, 46, 49, 50, 62, 63, 73, 75, 76, 77, 79, 82, 95, 98, 101, 102, 103, 105, 106, 107, 108, 109, 110, 111, 112, 113, 114, 115, 116, 118, 119, 120, 122, 123, 126, 127, 128, 130, 132, 133, 134
apoptosis, 47, 48, 107, 120
Armenia, 14, 45, 109, 130, 134
Asia, 24, 45, 76, 109, 118, 124, 134
assessment, 103, 118, 125, 126
asthma, 24, 43, 46, 76, 81, 82
Azerbaijan, 14, 45, 107

B

bacteria, 6, 47, 63, 113, 117, 125, 132
Bangladesh, 30, 31, 45, 102, 115
bark, 1, 4, 5, 6, 7, 8, 10, 14, 19, 27, 29, 31, 37, 38, 40, 46, 47, 50, 52, 53, 57, 58, 60, 63, 64, 70, 71, 72, 73, 79, 82, 84, 85, 86, 87, 88, 91, 92, 95, 96, 97, 98, 103, 108, 109, 110, 112, 114, 115, 119, 121, 122, 125, 127, 128, 129, 132, 134
base, 11, 58, 91
benefits, 2, 4, 15, 62, 76, 115, 117, 125
Betula utilis D. Don (Himalayan birch), v, 1, 2, 3, 4, 5, 6, 8, 101, 104, 114, 115, 117, 119, 120, 122, 127, 132, 134
betulinic acid, 5, 6, 64, 82, 84, 106, 130, 132
Bhutan, 3, 45, 76, 91
biodiversity, 2, 78, 111, 112, 123, 134
biological activities, 6, 9, 23, 24, 42, 47, 63, 80, 91, 106, 119, 128
bleeding, 75, 77, 81, 95
blood, 13, 14, 31, 48, 49, 60, 73, 75, 77, 79, 81, 104
blood pressure, 13, 60, 75, 77, 104
bones, 13, 43, 46, 79, 81

C

cancer, 5, 13, 15, 32, 38, 43, 48, 79, 82, 106, 109, 110, 112, 120, 125, 131

carbohydrate, 13, 24, 33, 34, 61, 84, 102, 129, 131
carbon, 92, 101, 112, 131
cardiovascular disease, 15, 79, 82
case study, 104, 106, 124, 134
catkins, 1, 10, 11, 43, 58
Caucasus, 103, 104, 105, 111
cell line, 5, 38, 48, 72, 107, 109, 116, 131
chemical, 5, 9, 52, 63, 80, 89, 91, 103, 106, 110, 111, 119, 124
China, 3, 27, 45, 46, 57, 60, 76, 81, 106, 109, 113, 118, 133, 134
chloroform, 7, 21, 37, 38, 40, 48, 50, 71, 73, 85, 86, 87, 88, 105
cholesterol, 13, 14, 28, 48, 106, 108, 112
classification, 103, 107, 117
climate, 2, 10, 12, 92, 117, 130
colon, 51, 62, 107, 118
commercial, 33, 44, 92, 123
community, 78, 103, 110, 111, 119, 122, 130
complications, 23, 42, 82
composition, 34, 92, 102, 106, 107, 108, 109, 110, 113, 115, 116, 118, 121, 122, 126
compounds, 5, 6, 15, 16, 22, 23, 24, 33, 34, 42, 48, 53, 54, 57, 63, 64, 79, 95, 96, 110, 114, 115, 117, 119, 121
conservation, 8, 106, 123, 130, 133
constituents, 1, 15, 102, 108, 111, 114, 115, 116, 121, 123, 124, 133
consumption, 13, 45, 60
copper, 28, 49, 80
Corylus avellana L. (Hazelnut), v, 9, 10, 11, 12, 13, 14, 15, 16, 17, 18, 19, 20, 21, 22, 102, 103, 104, 105, 106, 107, 108, 109, 110, 111, 112, 113, 116, 118, 119, 121, 122, 123, 124, 125, 127, 128, 129, 130, 133, 134
cough, 15, 24, 28, 82, 95
crop, 10, 23, 106, 107, 112, 116, 119, 125, 127
cultivars, 102, 103, 106, 111, 118, 127, 128
cultivation, 12, 58, 60, 63, 118, 130
cure, 15, 46, 80, 81, 82, 101, 125, 127, 134

D

derivatives, 5, 130, 133
diabetes, 13, 23, 24, 30, 41, 42, 43, 46, 48, 49, 57, 60, 80, 81, 105, 107, 110, 113, 114, 115, 119, 128
diarrhea, 14, 43, 46, 52, 75, 76, 77, 79, 95
diet, 13, 31, 33, 46, 62, 103, 113, 114, 117, 121, 132
digestion, 60, 95, 105
disease, 5, 13, 14, 15, 24, 28, 31, 32, 43, 46, 51, 57, 62, 63, 76, 79, 82, 89, 91, 95, 107, 111, 113, 118, 124, 132, 134
distribution, 10, 55, 63, 92, 104, 107, 111, 114
diversity, 103, 104, 117, 130, 132, 133
DNA, 50, 106, 109
drugs, 1, 5, 22, 55, 63, 89
drying, 46, 115, 128, 133

E

E. coli, 20, 39, 40, 71, 72, 85, 97
edible, 23, 27, 28, 46, 60, 92, 103, 107, 114, 122, 130
endangered, 2, 109, 111
energy, 13, 33, 34
enzymes, 27, 33, 52, 80, 103, 129
ester, 19, 53, 67, 83
ethanol, 7, 21, 37, 40, 41, 50, 51, 53, 71, 72, 73, 74, 122
ethyl acetate, 5, 7, 38, 40, 41, 72, 86, 87, 88, 131
Europe, 45, 57, 104, 134
extraction, 15, 29, 50, 80, 107, 109
extracts, 5, 6, 14, 23, 42, 47, 48, 49, 50, 51, 52, 62, 63, 89, 91, 101, 102, 105, 106, 110, 112, 115, 116, 117, 118, 120, 121, 122, 124, 125, 128, 130, 132, 133

F

fat, 13, 28, 34, 102, 103, 113, 114
fatty acids, 13, 33, 52, 61, 113
female, 1, 10, 11, 27, 43, 44, 49, 52, 58
fever, 14, 46, 60, 75, 77, 81, 82, 95

fibre, 13, 24, 28, 33, 34, 52, 61, 92
Ficus carica L. (Fig), v, 23, 24, 25, 26, 27, 28, 29, 31, 33, 34, 35, 37, 42, 101, 102, 103, 104, 105, 106, 107, 108, 109, 110, 111, 112, 113, 114, 115, 116, 117, 118, 119, 120, 121, 122, 123, 124, 125, 126, 127, 128, 129, 130, 131, 132, 133
fish, 61, 79, 81
flavonoids, 15, 23, 42, 49, 52, 63, 75, 77, 79, 80, 91, 107, 108, 109, 116, 119, 128, 131
flavonol, 104, 105, 108, 114, 129
flora, 78, 102, 111, 113
flowers, 1, 10, 11, 19, 27, 43, 44, 50, 58, 75, 76, 77, 78, 79, 81, 85, 87, 91, 101, 107, 110, 114, 119, 124, 127, 128, 129, 131
folk medicine, 4, 14, 46, 113
food, 9, 13, 28, 30, 33, 57, 59, 61, 62, 63, 81, 102, 103, 104, 105, 106, 107, 108, 109, 111, 112, 113, 114, 115, 116, 117, 118, 119, 120, 121, 122, 123, 124, 125, 126, 127, 128, 129, 130, 131, 132, 133, 134
frost, 92, 93, 124
fructose, 33, 61, 62
fruit, 9, 11, 13, 14, 15, 23, 24, 27, 28, 29, 30, 31, 32, 33, 36, 37, 38, 41, 46, 47, 49, 52, 57, 58, 60, 61, 62, 63, 71, 72, 73, 74, 75, 91, 92, 95, 103, 105, 108, 109, 111, 112, 113, 114, 115, 116, 117, 118, 119, 121, 123, 124, 125, 126, 128, 131, 132, 133
fungi, 6, 47, 125

G

genus, 1, 3, 9, 23, 24, 57, 75, 76, 103, 115
Georgia, 15, 104, 105, 116, 127
glucose, 18, 33, 48, 49, 54, 61, 62, 129
glucoside, 16, 17, 19, 35, 36, 53, 66, 67, 69, 70, 106
glycoside, 16, 82, 123, 134
growth, 4, 5, 30, 46, 47, 50, 59, 62, 93, 110, 117, 118

H

habitat, 2, 4, 8, 9, 76
hair, 46, 78, 81
headache, 75, 76, 77, 81, 82, 95
healing, 46, 98, 127, 133
health, 46, 60, 62, 76, 81, 102, 115, 117, 125, 128
heart disease, 5, 13, 76
height, 1, 43, 58, 76, 91
hexane, 5, 7, 20, 38, 40, 47, 72, 86, 87, 88
history, 45, 55, 112
human, 5, 6, 13, 15, 21, 23, 28, 33, 38, 42, 48, 61, 62, 73, 102, 107, 108, 109, 110, 116, 117, 124, 126, 128, 130, 131, 133, 134
human health, 13, 23, 42, 62, 110, 124

I

identification, 52, 113, 114, 128, 133
in vitro, 5, 48, 49, 101, 105, 106, 107, 108, 110, 119, 126
in vivo, 106, 107, 110, 124, 125
incompatibility, 2, 10, 111
India, 3, 10, 24, 29, 30, 31, 32, 45, 57, 60, 72, 75, 76, 78, 81, 82, 91, 103, 104, 105, 107, 109, 110, 111, 114, 116, 117, 118, 119, 120, 121, 122, 123, 124, 125, 126, 127, 130, 134, 151
industry, 59, 63, 99, 129
infection, 28, 46, 82, 106
inflammation, 28, 46, 109
inhibition, 52, 82, 98, 102, 110, 117, 120
Iran, 14, 26, 29, 31, 32, 45
Islam, 37, 41, 115, 119, 120
isolation, 48, 63, 104, 114, 125, 128, 133
Italy, 10, 23, 29, 30, 31

J

Japan, 60, 76, 118
jaundice, 60, 101, 125
Juglans regia L. (Walnut), v, 43, 44, 45, 46, 47, 48, 49, 50, 51, 52, 53, 54, 55, 111

Index

K

kaempferol, 16, 17, 18, 19, 35, 53, 68
kernels, 9, 11, 13, 14, 16, 17, 18, 19, 20, 21, 46, 102, 106, 107, 109, 110, 113, 124, 127, 128, 134
kidney, 4, 52, 113

L

Lactobacillus, 20, 62, 119
LDL, 48, 49, 109, 114
lead, 48, 63, 80
leaves, 1, 2, 4, 5, 6, 7, 9, 10, 12, 14, 15, 16, 17, 19, 20, 21, 23, 27, 28, 29, 30, 31, 32, 35, 37, 38, 39, 40, 41, 42, 43, 46, 47, 48, 49, 50, 51, 52, 53, 58, 59, 60, 62, 63, 67, 69, 70, 71, 72, 73, 74, 75, 76, 77, 78, 79, 80, 81, 82, 83, 85, 86, 87, 88, 91, 92, 93, 95, 96, 97, 98, 101, 102, 105, 106, 107, 108, 109, 110, 111, 113, 114, 115, 116, 117, 118, 119, 120, 121, 122, 123, 124, 125, 126, 127, 128, 129, 130, 131, 132, 133, 134
light, 2, 12, 93, 105
linoleic acid, 33, 35, 61, 83
lipids, 9, 33, 34, 61, 129
liver, 5, 14, 21, 23, 24, 28, 38, 42, 46, 52, 60, 82, 112

M

Malaysia, 26, 30, 32, 76
male, 1, 7, 10, 11, 27, 37, 41, 43, 44, 49, 52, 53, 58, 71, 73, 87, 88, 98, 126
management, 75, 92, 119, 123, 124
maturity, 2, 43, 44, 118
medication, 4, 75, 95
medicinal, 1, 14, 22, 23, 28, 42, 43, 57, 60, 75, 77, 79, 89, 91, 92, 99, 101, 102, 104, 105, 106, 107, 108, 109, 110, 111, 112, 113, 115, 116, 117, 118, 120, 122, 123, 124, 126, 127, 128, 129, 130, 131, 132, 134
medicine, 1, 4, 8, 14, 15, 22, 23, 24, 28, 42, 43, 46, 57, 60, 75, 78, 79, 91, 95, 101, 104, 109, 113, 116, 119, 124, 126, 129, 132, 133, 151
Mediterranean, 10, 23, 24, 28, 31, 45, 105, 131, 134
mellitus, 43, 107, 113
metabolites, 9, 15, 34, 63, 91, 133
methanol, 7, 21, 37, 38, 39, 40, 41, 48, 50, 71, 72, 73, 74, 85, 86, 97, 106, 107, 113, 126
mice, 41, 51, 52, 71, 73, 87, 88, 98, 101, 109, 121, 133
minerals, 13, 24, 33, 34, 52, 61, 80, 121, 128
models, 49, 50, 51, 79, 126, 131
morphology, 23, 43, 48, 107, 122
Morus alba L. (Mulberry), v, 57, 58, 59, 60, 61, 62, 63, 64, 71, 101, 102, 103, 105, 106, 107, 108, 109, 111, 112, 113, 114, 115, 116, 117, 118, 119, 120, 121, 123, 124, 125, 126, 128, 129, 130, 131, 132, 133, 134
Moscow, 104, 109, 111, 113, 130
Myanmar, 76, 81, 122

N

Nepal, 3, 30, 45, 46, 76, 91, 92, 102, 104, 106, 107, 108, 110, 111, 112, 113, 115, 117, 119, 121, 122, 123, 126, 127, 132
nitric oxide, 51, 101, 107, 125, 133
nutrients, 2, 12, 33, 80, 116, 126, 132
nutrition, 13, 61, 103, 106, 107, 109, 110, 128, 129, 130, 132, 133
nuts, 9, 10, 11, 13, 14, 17, 20, 43, 44, 45, 46, 53, 102, 106, 107, 111, 116

O

obesity, 63, 73, 91, 95, 98, 119, 121, 129
oil, 6, 13, 14, 15, 28, 49, 61, 92, 95, 103, 109, 115, 128
oleic acid, 33, 35, 61

P

pain, 14, 24, 28, 29, 31, 46, 81

Pakistan, 3, 26, 27, 29, 30, 31, 32, 45, 60, 76, 82, 101, 111, 112, 113, 115, 118, 119, 124, 127, 134
petroleum, 7, 37, 38, 39, 41, 47, 50, 71, 73, 85, 97
pharmacology, 9, 43, 55, 63, 75, 91, 103, 105, 128
phenolic compounds, 15, 24, 34, 109, 115, 123, 126, 128
phytochemicals, 9, 15, 34, 35, 40, 43, 52, 57, 63, 75, 77, 80, 83, 89, 91, 95, 99, 101, 102, 105, 106, 108, 111, 112, 114, 115, 116, 117, 118, 120, 126, 127, 132
plants, 8, 10, 44, 52, 59, 60, 75, 76, 83, 92, 93, 101, 102, 103, 107, 110, 111, 112, 113, 114, 115, 117, 119, 120, 121, 122, 124, 125, 126, 127, 128, 129, 130, 132, 134
pneumonia, 20, 39, 47, 85, 86, 97
pollen, 1, 10, 11, 17, 44, 112, 122, 129
polyphenols, 23, 24, 42, 52, 111, 116
population, 2, 8, 75, 114, 117, 125
prevention, 23, 28, 42, 62, 63, 76, 106, 120
proliferation, 48, 112, 125, 131, 134
protection, 1, 2, 8, 46
protein, 13, 27, 34, 52, 59, 60, 61, 84, 114, 133
Pseudomonas aeruginosa, 6, 7, 20, 39, 71, 85, 97
pulp, 37, 40, 60, 102

Q

quantification, 115, 128, 129, 133
quercetin, 16, 17, 19, 35, 49, 53, 68, 109, 127, 129

R

receptor, 51, 109, 121
regeneration, 117, 119, 120, 126, 129
religious, 1, 4, 8
requirements, 13, 28, 114, 133
researchers, 24, 52, 80, 95
resources, 78, 99, 103, 108, 112

Rhododendron arboreum Sm., v, 4, 75, 76, 77, 78, 79, 80, 81, 82, 83, 85, 89, 101, 104, 106, 107, 110, 111, 112, 113, 114, 115, 116, 117, 118, 119, 120, 121, 122, 123, 124, 125, 126, 127, 128, 129, 130, 131, 133, 134
risk, 13, 15, 118
roots, 5, 6, 7, 15, 23, 27, 28, 30, 31, 46, 48, 52, 63, 64, 70, 71, 72, 73, 75, 77, 79, 82, 84, 85, 87, 96, 108, 109, 112, 121, 127, 128, 132, 134

S

Salmonella, 7, 20, 39, 47, 71, 85, 97
seeds, 2, 11, 15, 27, 28, 45, 46, 53, 54, 61, 78, 92, 93, 96, 97, 123, 127, 131
Shorea robusta Gaertn. f. (Sal trees), v, 91, 92, 93, 94, 95, 96, 97, 99, 102, 103, 106, 107, 108, 110, 112, 114, 115, 118, 119, 120, 122, 123, 124, 125, 126, 127, 128, 129, 130, 131, 132, 133
silkworms, 57, 59, 117
skin, 9, 15, 17, 18, 19, 20, 24, 29, 33, 43, 46, 52, 63, 82, 107, 119, 123
species, 1, 2, 3, 4, 8, 9, 10, 15, 23, 24, 27, 43, 44, 45, 47, 55, 57, 58, 63, 75, 76, 78, 79, 80, 91, 92, 93, 95, 101, 104, 105, 107, 110, 113, 115, 117, 118, 125, 127, 129, 134
Sri Lanka, 76, 77, 119
state, 10, 76, 122
steroids, 52, 80, 82
stomach, 46, 51, 112
sugar, 24, 28, 30, 49, 60, 61, 62, 79, 82, 84, 95, 102, 118, 128
supplementation, 33, 108, 112, 128

T

tannins, 52, 75, 77, 80, 82
techniques, 15, 80, 104
temperature, 12, 79, 93, 115
therapeutic use, 1, 8, 60
tonic, 14, 28, 29, 30, 79, 81, 105
toxicity, 42, 52, 129, 132

trade, 11, 28, 45
traditional medicine, 14, 15, 22, 23, 42, 43, 46, 79, 91, 101
treatment, 14, 15, 22, 23, 28, 29, 30, 42, 46, 48, 49, 52, 75, 82, 91, 95, 101, 104, 106, 107, 113, 120, 134
tree, 1, 2, 3, 4, 8, 9, 10, 14, 15, 16, 20, 23, 26, 27, 33, 43, 44, 45, 46, 58, 60, 75, 76, 78, 91, 92, 93, 94, 102, 104, 106, 109, 112, 113, 115, 116, 117, 118, 121, 122, 123, 124, 125, 126, 129, 130, 134, 151
Turkey, 10, 12, 23, 27, 29, 30, 32, 45, 60, 103, 105, 111, 116, 117, 130

U

ulcer, 24, 31, 51, 126
United States (USA), 108, 111, 121, 123
USSR, 109, 111, 130

V

varieties, 24, 47, 57, 60, 95, 110, 112, 116, 118, 121, 122, 123, 128
vitamin E, 13, 52, 61, 83, 107, 131
vitamins, 13, 14, 24, 28, 34, 52, 53, 61, 83, 104, 107, 112, 118, 131

W

water, 2, 7, 8, 14, 20, 21, 29, 30, 32, 38, 41, 50, 71, 72, 73, 74, 79, 98, 112, 133
watershed, 76, 126, 129
wood, 4, 7, 8, 10, 30, 60, 76, 82, 92, 104, 108, 124, 127
World Health Organization (WHO), 61, 75, 114, 133
wound healing, 91, 95, 107, 120

Y

yield, 10, 13, 93

Authors' Contact Information

Dr. Mohammad Saleem Wani
Lecturer
Govt Degree College for Women Sopore
Jammu and Kashmir, India
saleemwani806@gmail.com

Dr. Younas Rasheed Tantray
Research Assistant
Department of Botany
Central Ayurveda Research Institute, Jhansi,
Uttar Pradesh, India
younasrasheed53@gmail.com

Dr. Afaq Majid Wani
Head and Associate Professor
Department of Forest Biology, Tree Improvement
and Wildlife Sciences, College of Forestry,
Sam Higginbottom University of Agriculture, Technology and Sciences
Pragyaraj (Uttar Pradesh), India
afaqtree@gmail.com